T0091912

Robotics

International Series on
INTELLIGENT SYSTEMS, CONTROL, AND AUTOMATION:
SCIENCE AND ENGINEERING

VOLUME 43

For other titles published in this series, go to
http://www.springer.com/series/6259

Tadej Bajd • Matjaž Mihelj • Jadran Lenarčič
Aleš Stanovnik • Marko Munih

Robotics

Professor Tadej Bajd
University of Ljubljana
Fac. Electrical Engineering
Tržaška 25
SI-1000 Ljubljana
Slovenia
tadej.bajd@robo.fe.uni-lj.si

Professor Aleš Stanovnik
University of Ljubljana
Fac. Electrical Engineering
Tržaška 25
SI-1000 Ljubljana
Slovenia
ales.stanovnik@ijs.si

Professor Matjaž Mihelj
University of Ljubljana
Fac. Electrical Engineering
Tržaška 25
SI-1000 Ljubljana
Slovenia
matjaz.mihelj@robo.fe.uni-lj.si

Professor Marko Munih
University of Ljubljana
Fac. Electrical Engineering
Tržaška 25
SI-1000 Ljubljana
Slovenia
marko.munih@robo.fe.uni-lj.si

Professor Jadran Lenarčič
University of Ljubljana
Inst. J. Stefan
SI-1111 Ljubljana
Slovenia

ISBN 978-90-481-3775-6 e-ISBN 978-90-481-3776-3
DOI 10.1007/978-90-481-3776-3
Springer Dordrecht Heidelberg London New York

Library of Congress Control Number: 2010920152

Cover design: eStudio Calamar S.L.

Printed on acid-free paper

Springer is part of Springer Science+Business Media (www.springer.com)

Preface

The word "robot" does not originate from a science or engineering vocabulary. It was first used in the Czech drama R.U.R. (Rossum's Universal Robots) written by Karel Čapek and was first played in Prague in 1921 (the word itself was invented by his brother Josef). In the drama the "robot" is an artificial human being which is a brilliant worker, deprived of all unnecessary qualities: feelings, creativity and capacity for feeling pain. In the prologue of the drama the following "definition" of robots is given: *Robots are not people (Roboti nejsou lidé). They are mechanically more perfect than we are, they have an astounding intellectual capacity, but they have no soul. The creation of an engineer is technically more refined than the product of nature.*

The textbook "Robotics" evolved through more than 10 years of teaching robotics at the Faculty of Electrical Engineering, of the University of Ljubljana, Slovenia. The way of presenting the rather demanding subject was successfully tested with several generations of undergraduate students.

The major feature of the book is its simplicity. The basic characteristics of industrial robot mechanisms are presented in the introduction. The position, orientation and displacement of an object are described by homogenous transformation matrices. These matrices, which are the basis for any analysis of robot mechanisms, are introduced through simple geometrical reasoning. Geometrical models of the robot mechanism are explained with the help of an original and friendly vector description. Robot kinematics and dynamics are introduced via a mechanism with only two rotational degrees of freedom, which is however an important part of the most popular industrial SCARA and anthropomorphic robot structures. The presentation of robot dynamics is based on only the knowledge of Newton's law. The robot workspace plays an important role in selecting an appropriate robot for the task planned. Robot sensors and robot trajectory planning are presented. Basic control schemes, resulting in either the desired end-effector trajectory or in the force between the robot and its environment, are also explained. Robot grippers and feeding devices are described together with the planning of robot assembly. The chapter on standardization and measurement of accuracy and repeatability is of interest for users of industrial robots. The textbook is supplemented with a short English–German–French robotic vocabulary.

The book requires minimal advance knowledge of mathematics and physics. Therefore it is appropriate for students of engineering schools (electrical, mechanical, computer, civil) or first-level students according to the two-level Bologna program. It could be of interest also for engineers who did not study robotics, but encounter robots in their working environment and wish to acquire some basic knowledge in a simple and fast manner.

The authors acknowledge the precious help of Professor Robert Riener from ETH, Zürich and Professor Christine Azevedo from LIRMM, Montpellier in preparation of the English–German–French robotic vocabulary.

Ljubljana
July 2009

Tadej Bajd
Matjaž Mihelj

Contents

Chapter 1
Introduction

It is appropriate to begin the textbook on robotics with the definition of the industrial robot manipulator as given by the ISO 8373 standard. An industrial robot manipulator is a feedback controlled, reprogrammable, multipurpose system. It is programmable in three or more degrees of freedom. Robot manipulators are used in processes of industrial automation.

The standard stresses feedback control of industrial robots. Robotic mechanisms are actuated by electric and hydraulic motors. Important component parts of any robotic system are sensors. Here, we distinguish between internal and external sensors. Internal sensors assess position and velocity of robot segments and are placed into robotic joints. Among external sensors, the most important are the sensor of contact forces and the robot vision sensors. The aim of the robot control system is to guide the robot end-point with respect to the desired trajectory determined by the user and with respect to information received from the sensors.

In modern industrial production, there are no large stocks of either materials or of products. We say that the production process runs just in time. As a consequence, it may happen that different types of a certain product find themselves on the same production line during the same day. The problem, which is most inconvenient for fixed automation, can be efficiently solved by the use of industrial robotic manipulators. Reprogrammable robots allow us to switch from the production of one type of product to another similar one simply by touching a push-button.

Furthermore, the ISO standard definition characterizes the robot manipulator as a multipurpose mechanism. The robot mechanism is a crude imitation of the human arm. In the same way as we use our arm for both precise and heavy work, we are trying to apply the same robot manipulator to different tasks. This is even more important in view of the economic life span of an industrial robot, which is rather long (12–16 years). It can therefore happen, that we had acquired a robot manipulator for welding purposes, while after certain period of time the robot will be used for a pick and place task.

T. Bajd et al., *Robotics*, Intelligent Systems, Control and Automation: Science and Engineering 43, DOI 10.1007/978-90-481-3776-3_1,

The last property, expressed in the definition, describes the robot as a mechanism, which is programmable in three or more degrees of freedom. As this is the most characteristic property of an industrial robot, we shall examine more closely the meaning of a degree of freedom.

1.1 Degree of freedom

To begin with, we will introduce the degree of freedom having in mind an infinitesimal mass particle. In this case the number of degrees of freedom is defined as the number of independent coordinates (not including time) which are necessary for the complete description of the position of a mass particle.

A particle moving along a line (infinitesimally small ball on a wire) is a system with one degree of freedom. A pendulum with a rigid segment, which is swinging in a plane, is also a system with one degree of freedom (Figure 1.1). In the first example the position of the particle can be described with the distance, while in the second case it is described with the angle of rotation.

A mass particle moving on a plane has two degrees of freedom (Figure 1.2). The position of the particle can be described with two cartesian coordinates x and y. The double pendulum with rigid segments, swinging in a plane, is also a system with two degrees of freedom. The position of the mass particle is described by two angles. A mass particle in space has three degrees of freedom. Usually its position is expressed by three rectangular coordinates x, y and z. An example of a simple mechanical system with three degrees of freedom is a double pendulum where one segment is represented by an elastic spring and the other by a rigid rod. Also in this case the pendulum is swinging in a plane.

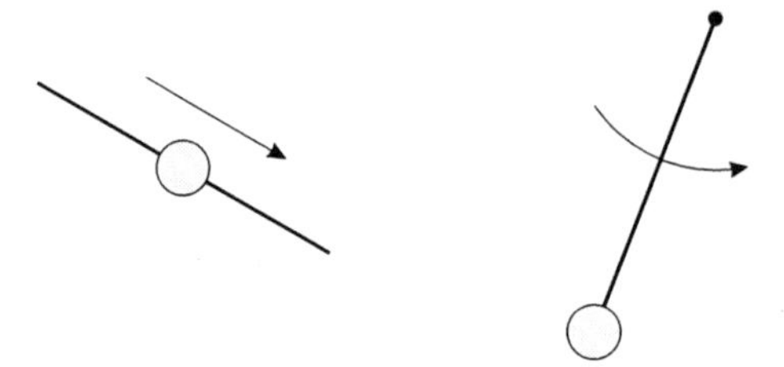

Fig. 1.1 Two examples of systems with one degree of freedom: mass particle on a wire (left) and rigid pendulum in a plane (right)

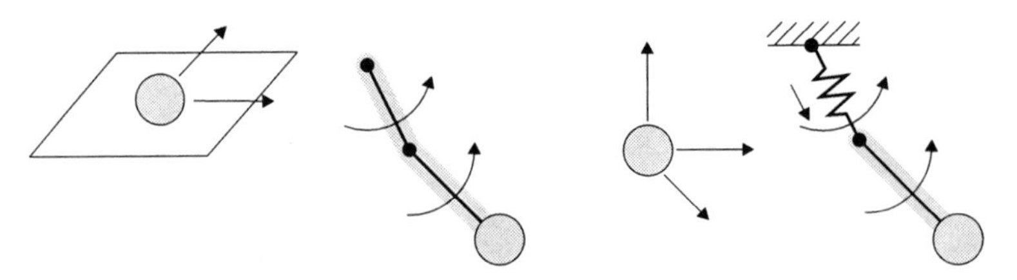

Fig. 1.2 Examples with two (left) and three degrees of freedom (right)

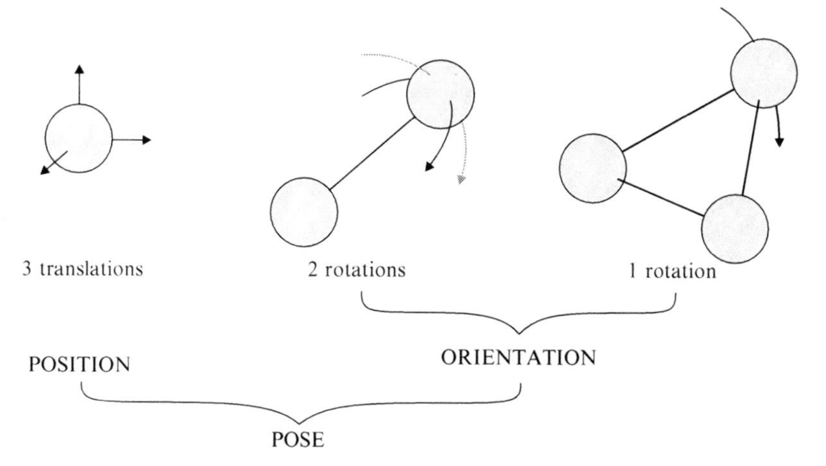

Fig. 1.3 Degrees of freedom of a rigid body

In robotics we are not interested in mass particles but rather in rigid bodies, which are either robot segments or objects manipulated by the industrial robot. The simplest rigid body consists of three mass particles (Figure 1.3). We already know that a single mass particle has three degrees of freedom, described by three rectangular displacements along a line called translations (T). We add another mass particle to the first one in such a way that there is constant distance between them. The second particle is restricted to move on the surface of a sphere surrounding the first particle. Its position on the sphere can be described by two circles reminding us of meridians and latitudes on a globe. The displacement along a circular line is called rotation (R). The third mass particle is added in such a way that the distances with respect to the first two particles are kept constant. In this way the third particle may move along a circle, a kind of equator, around the axis determined by the first two particles. A rigid body therefore has six degrees of freedom: three translations and three rotations. The first three degrees of freedom describe the position of the body, while the other three degrees of freedom determine its orientation. The term pose is used to include both position and orientation.

1.2 Robot manipulator

The robot manipulator consists of a robot arm, wrist, and gripper (Figure 1.4). The task of the robot manipulator is to place an object grasped by the gripper into an arbitrary pose. In this way also the industrial robot needs to have six degrees of freedom. The segments of the robot arm are relatively long. The task of the robot arm is to provide the desired position of the robot end point. The segments of the robot wrist are rather short. The task of the robot wrist is to enable the required orientation of the object grasped by the robot gripper.

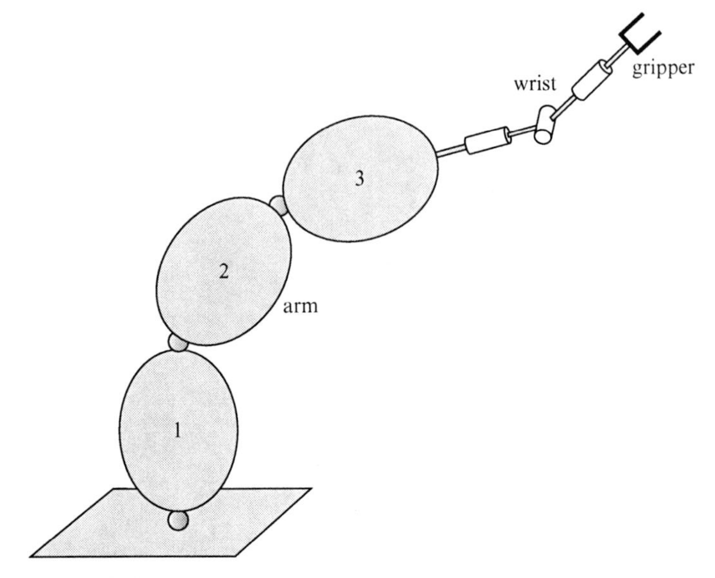

Fig. 1.4 Robot manipulator

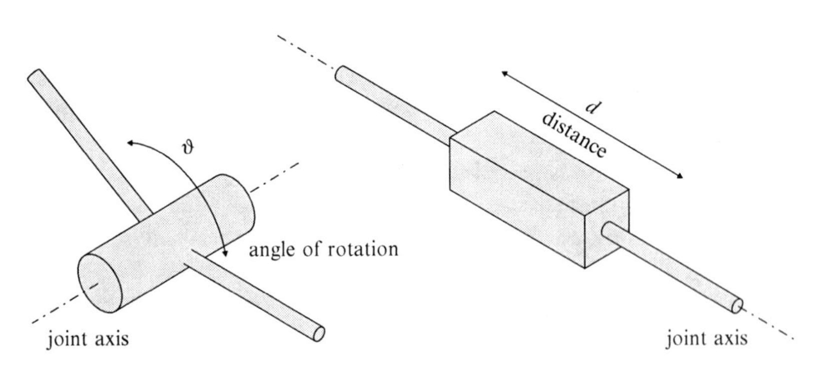

Fig. 1.5 Rotational (left) and translational (right) robot joint

The robot arm is a serial chain of three rigid bodies called robot segments. Two neighbor segments of a robot manipulator are connected through a robot joint. The joint decreases the number of degrees of freedom which occur between two neighbor segments. The robot joints have only one degree of freedom and are either translational or rotational (Figure 1.5).

The rotational joint has the form of a hinge and limits the motion of two neighbor segments to rotation around the joint axis. The relative position of the segments is given by the angle of rotation around the joint axis. In robotics the joint angles are denoted by the Greek letter ϑ. In simplified robotic models the rotational joint is represented by a cylinder. The translational joint restricts the movement of two neighboring segments to translation. The relative position between the two segments is measured as a distance. The symbol of the translational joint is a prism, while the distance is denoted by the letter d.

1.3 Robot arms

We have seen that robot joints are either translational or rotational. Robot arms have another important property. The axes of two neighboring joints are either parallel or perpendicular. As the robot arm has only three degrees of freedom, there exist a limited number of possible combinations resulting all together in 36 different structures of robot arms. Among them only 12 are functionally different. On the market we find 5 commercially available structures of robot arms: anthropomorphic, spherical, SCARA, cylindrical, and cartesian.

The anthropomorphic robot arm (Figure 1.6) has all three joints of the rotational type (RRR). Among the robot arms it resembles the human arm to the largest extent. The second joint axis is perpendicular to the first one, while the third joint axis is parallel to the second one. The workspace of the anthropomorphic robot arm, encompassing all the points that can be reached by the robot end point, has a spherical shape.

The spherical robot arm (Figure 1.7) has two rotational and one translational degree of freedom (RRT). The second joint axis is perpendicular to the first one and the third axis is perpendicular to the second one. The workspace of the robot arm has a spherical shape as in the case of the anthropomorphic robot arm.

The SCARA (Selective Compliant Articulated Robot for Assembly) robot arm appeared relatively late in the development of industrial robotics (Figure 1.8). It is predominantly aimed for industrial processes of assembly. Two joints are rotational and one is translational (RRT). The axes of all three joints are parallel. The workspace of SCARA robot arm is of cylindrical shape.

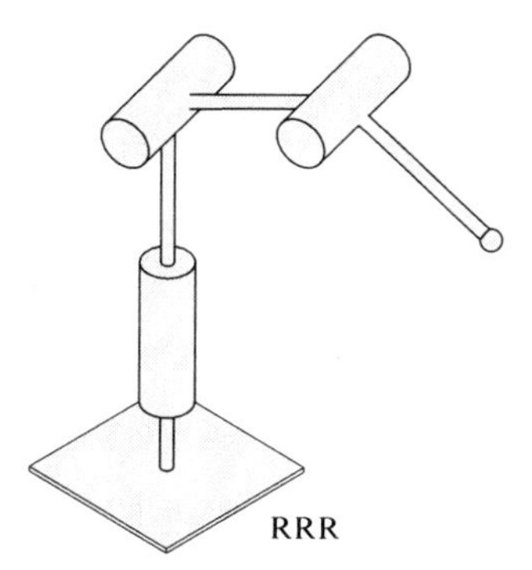

Fig. 1.6 Anthropomorphic robot arm

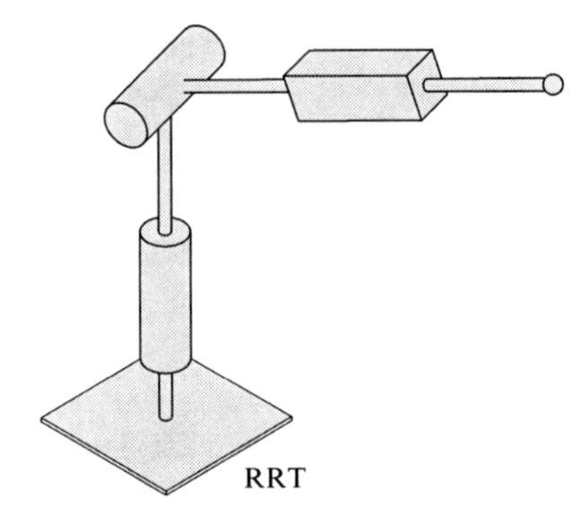

Fig. 1.7 Spherical robot arm

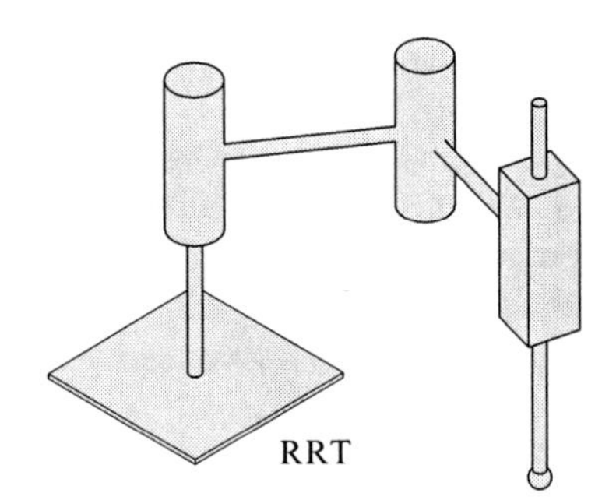

Fig. 1.8 SCARA robot arm

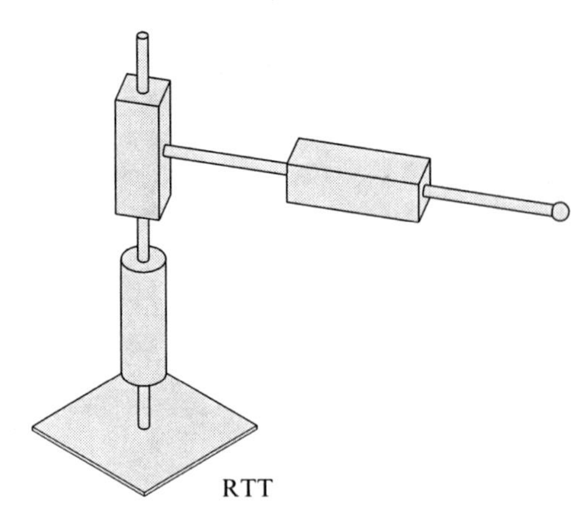

Fig. 1.9 Cylindrical robot arm

The cylindrical shape of the workspace is even more evident with the cylindrical robot arm (Figure 1.9). This robot has one rotational and two translational degrees of freedom (RTT). The axis of the second joint is parallel to the first axis, while the third joint axis is perpendicular to the second one.

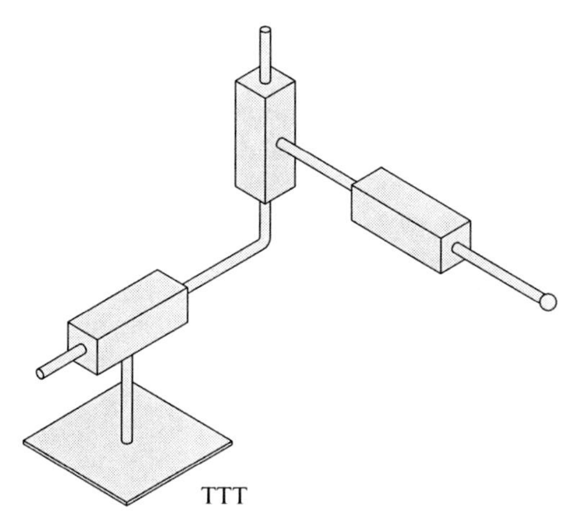

Fig. 1.10 Cartesian robot arm

The cartesian robot arm (Figure 1.10) has all three joints of the translational type (TTT). The joint axes are perpendicular one to another. Cartesian robot arms are known for high accuracy, while the special structure of gantry robots is suitable for manipulation of heavy objects. The workspace of the cartesian robot arm is a prism.

1.4 Robot manipulators in industrial environment

Today we encounter the largest number of industrial robot manipulators in the car industry. They are predominantly used for welding. The ratio of human workers and robots in the car industry is 6:1. In many cases the industrial robots are used for tasks where the robot gripper moves objects from point to point. Such examples are found in the process of palletizing, this means putting in order component parts or products for the purposes of feeding a machine or packaging. Industrial robots are often used in aggressive or dangerous environments, such as spray painting. Robot manipulators are increasingly entering the area of industrial assembly of component parts into a functional system. Robot manipulators are not encountered only in industrial environments. They are of more and more interest in medicine. We find them in surgical applications (hip joint replacement) or in rehabilitation (training of paralyzed extremity after stroke). Special examples of robot manipulators are telemanipulators. These are robots which are controlled by a human operator. They are used in dangerous environments or distant places.

Different from robot manipulators, which represent the main interest of this textbook, are mobile robots. They are either wheeled or legged. The wheeled mobile robots are used on even terrain. As their pose can be described by only three degrees of freedom, they are simpler to control than robot manipulators. Their strength is in the use of robot vision and other sensors assessing distance or contact with objects in the environment. Today they are mainly used for cleaning and mowing purposes. The biologically inspired legged mobile robots usually have six legs and are used on uneven terrain. An efficient representative is the forestry robot which is also capable of cutting trees. A counterpart to industrial robotics is the so called service robotics where robots are used to help people (predominantly graying population) in daily activities. The most advanced examples are humanoid robots capable of biped locomotion. Robots in the air and in the sea are no surprise. They are used for observation of distant terrains or for ocean studies. Sophisticated robotic toys are appreciated by children. Finally, robots are replacing humans also at such a noble occupation as the arts. They are dancing, playing musical instruments, and even painting.

Chapter 2
Homogenous transformation matrices

2.1 Translational transformation

In the introductory chapter we have seen that robots have either translational or rotational joints. We therefore need a unified mathematical description of translational and rotational displacements. The translational displacement **d**, given by the vector

$$\mathbf{d} = a\mathbf{i} + b\mathbf{j} + c\mathbf{k}, \tag{2.1}$$

can be described also by the following homogenous transformation matrix **H**

$$\mathbf{H} = Trans(a,b,c) = \begin{bmatrix} 1 & 0 & 0 & a \\ 0 & 1 & 0 & b \\ 0 & 0 & 1 & c \\ 0 & 0 & 0 & 1 \end{bmatrix}. \tag{2.2}$$

When using homogenous transformation matrices an arbitrary vector has the following 4×1 form

$$\mathbf{q} = \begin{bmatrix} x \\ y \\ z \\ 1 \end{bmatrix} = \begin{bmatrix} x & y & z & 1 \end{bmatrix}^T. \tag{2.3}$$

A translational displacement of vector **q** for a distance **d** is obtained by multiplying the vector **q** with the matrix **H**

$$\mathbf{v} = \begin{bmatrix} 1 & 0 & 0 & a \\ 0 & 1 & 0 & b \\ 0 & 0 & 1 & c \\ 0 & 0 & 0 & 1 \end{bmatrix} \begin{bmatrix} x \\ y \\ z \\ 1 \end{bmatrix} = \begin{bmatrix} x+a \\ y+b \\ z+c \\ 1 \end{bmatrix}. \tag{2.4}$$

The translation, which is presented by multiplication with a homogenous matrix, is equivalent to the sum of vectors **q** and **d**

$$\mathbf{v} = \mathbf{q} + \mathbf{d} = (x\mathbf{i} + y\mathbf{j} + z\mathbf{k}) + (a\mathbf{i} + b\mathbf{j} + c\mathbf{k}) = (x+a)\mathbf{i} + (y+b)\mathbf{j} + (z+c)\mathbf{k}. \tag{2.5}$$

T. Bajd et al., *Robotics*, Intelligent Systems, Control and Automation: Science and Engineering 43, DOI 10.1007/978-90-481-3776-3_2,

In a simple example, the vector $2\mathbf{i}+3\mathbf{j}+2\mathbf{k}$ is translationally displaced for the distance $4\mathbf{i}-3\mathbf{j}+7\mathbf{k}$

$$\mathbf{v} = \begin{bmatrix} 1 & 0 & 0 & 4 \\ 0 & 1 & 0 & -3 \\ 0 & 0 & 1 & 7 \\ 0 & 0 & 0 & 1 \end{bmatrix} \begin{bmatrix} 2 \\ 3 \\ 2 \\ 1 \end{bmatrix} = \begin{bmatrix} 6 \\ 0 \\ 9 \\ 1 \end{bmatrix}.$$

The same result is obtained by adding the two vectors.

2.2 Rotational transformation

Rotational displacements will be described in a right-handed rectangular coordinate frame, where the rotations around the three axes, as shown in Figure 2.1, are considered as positive. Positive rotations around the selected axis are counter-clockwise when looking from the positive end of the axis towards the origin of the frame O. The positive rotation can be described also by the so called right hand rule, where the thumb is directed along the axis towards its positive end, while the fingers show the positive direction of the rotational displacement. The direction of running of athletes on a stadium is also an example of a positive rotation.

Let us first take a closer look at the rotation around the x axis. The coordinate frame x', y', z' shown in Figure 2.2 was obtained by rotating the reference frame x, y, z in the positive direction around the x axis for the angle α. The axes x and x' are collinear.

The rotational displacement is also described by a homogenous transformation matrix. The first three rows of the transformation matrix correspond to the x, y and z axes of the reference frame, while the first three columns refer to the x', y' and z'

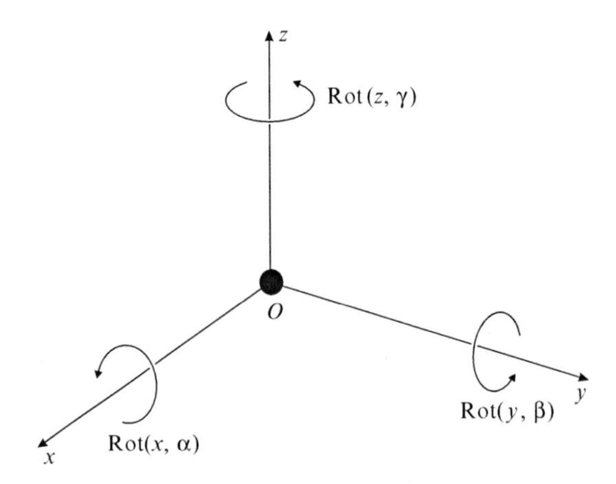

Fig. 2.1 Right-hand rectangular frame with positive rotations

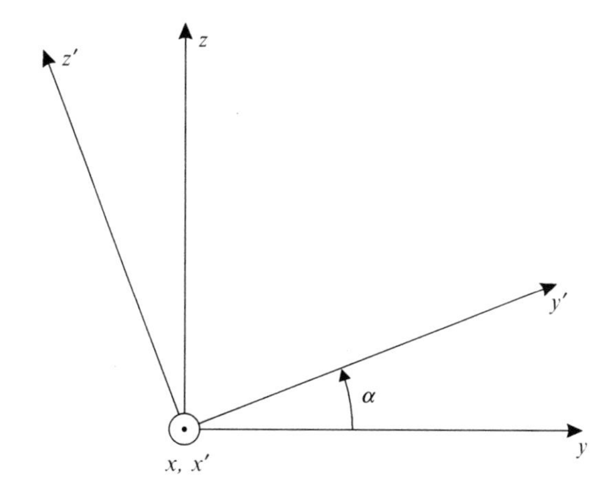

Fig. 2.2 Rotation around x axis

axes of the rotated frame. The upper left nine elements of the matrix **H** represent the 3×3 rotation matrix. The elements of the rotation matrix are cosines of the angles between the axes given by the corresponding column and row

$$Rot(x,\alpha) = \begin{array}{c} \begin{matrix} x' & \quad\quad y' & \quad\quad\quad z' & \end{matrix} \\ \begin{bmatrix} \cos 0^\circ & \cos 90^\circ & \cos 90^\circ & 0 \\ \cos 90^\circ & \cos\alpha & \cos(90^\circ+\alpha) & 0 \\ \cos 90^\circ & \cos(90^\circ-\alpha) & \cos\alpha & 0 \\ 0 & 0 & 0 & 1 \end{bmatrix} \end{array} \begin{matrix} x \\ y \\ z \\ \end{matrix}$$

$$= \begin{bmatrix} 1 & 0 & 0 & 0 \\ 0 & \cos\alpha & -\sin\alpha & 0 \\ 0 & \sin\alpha & \cos\alpha & 0 \\ 0 & 0 & 0 & 1 \end{bmatrix}. \tag{2.6}$$

The angle between the x' and the x axes is 0°, therefore we have $\cos 0^\circ$ in the intersection of the x' column and the x row. The angle between the x' and the y axes is 90°, we put $\cos 90^\circ$ in the corresponding intersection. The angle between the y' and the y axes is α, the corresponding matrix element is $\cos\alpha$.

To become more familiar with rotation matrices, we shall derive the matrix describing a rotation around the y axis by using Figure 2.3. Now the collinear axes are y and y'

$$y = y'. \tag{2.7}$$

By considering the similarity of triangles in Figure 2.3, it is not difficult to derive the following two equations

$$\begin{aligned} x &= x'\cos\beta + z'\sin\beta \\ z &= -x'\sin\beta + z'\cos\beta. \end{aligned} \tag{2.8}$$

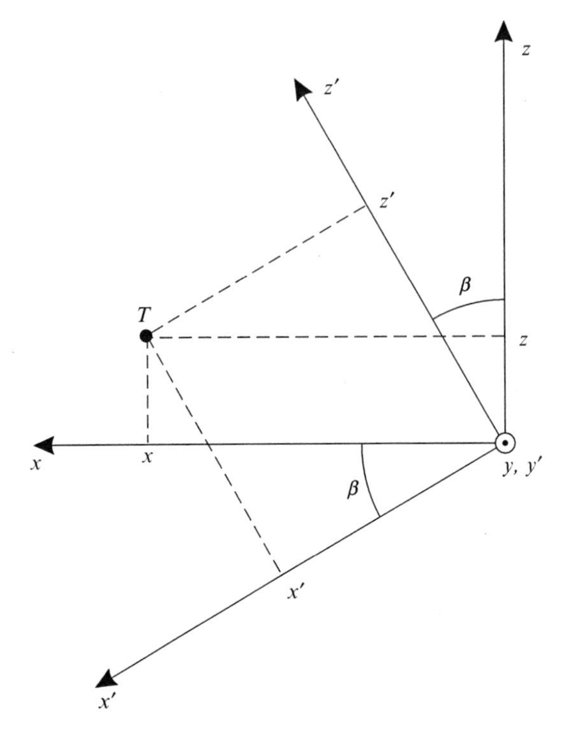

Fig. 2.3 Rotation around y axis

All three equations (2.7) and (2.8) can be rewritten in the matrix form

$$Rot(y,\beta) = \begin{array}{c} \begin{matrix} x' & y' & z' & \end{matrix} \\ \left[\begin{matrix} \cos\beta & 0 & \sin\beta & 0 \\ 0 & 1 & 0 & 0 \\ -\sin\beta & 0 & \cos\beta & 0 \\ 0 & 0 & 0 & 1 \end{matrix}\right] \begin{matrix} x \\ y \\ z \\ \end{matrix} \end{array}. \tag{2.9}$$

The rotation around the z axis is described by the following homogenous transformation matrix

$$Rot(z,\gamma) = \begin{bmatrix} \cos\gamma & -\sin\gamma & 0 & 0 \\ \sin\gamma & \cos\gamma & 0 & 0 \\ 0 & 0 & 1 & 0 \\ 0 & 0 & 0 & 1 \end{bmatrix}. \tag{2.10}$$

In a simple numerical example we wish to determine the vector $\mathbf{w}$ which is obtained by rotating the vector $\mathbf{u} = 7\mathbf{i} + 3\mathbf{j} + 0\mathbf{k}$ for 90° in the counter clockwise i.e. positive direction around the z axis. As $\cos 90^\circ = 0$ and $\sin 90^\circ = 1$, it is not difficult to determine the matrix describing $Rot(z, 90^\circ)$ and multiplying it by the vector $\mathbf{u}$

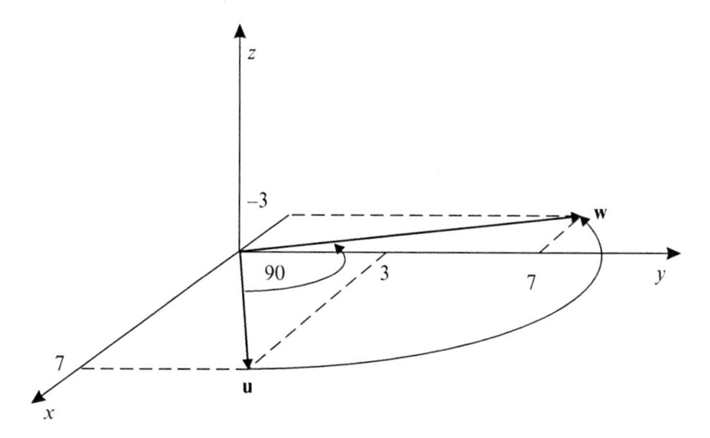

Fig. 2.4 Example of rotational transformation

$$\mathbf{w} = \begin{bmatrix} 0 & -1 & 0 & 0 \\ 1 & 0 & 0 & 0 \\ 0 & 0 & 1 & 0 \\ 0 & 0 & 0 & 1 \end{bmatrix} \begin{bmatrix} 7 \\ 3 \\ 0 \\ 1 \end{bmatrix} = \begin{bmatrix} -3 \\ 7 \\ 0 \\ 1 \end{bmatrix}.$$

The graphical presentation of rotating the vector **u** around the z axis is shown in Figure 2.4.

2.3 Pose and displacement

In the previous section we have learned how a point is translated or rotated around the axes of the cartesian frame. In continuation we shall be interested in displacements of objects. We can always attach a coordinate frame to a rigid object under consideration. In this section we shall deal with the pose and the displacement of rectangular frames. We shall see that a homogenous transformation matrix describes either the pose of a frame with respect to a reference frame, or it represents the displacement of a frame into a new pose. In the first case the upper left 3×3 matrix represents the orientation of the object, while the right-hand 3×1 column describes its position (e.g. the position of its center of mass). The last row of the homogenous transformation matrix will be always represented by 0, 0, 0, 1. In the case of object displacement, the upper left matrix corresponds to rotation and the right-hand column corresponds to translation of the object. We shall examine both cases through simple examples. Let us first clear up the meaning of the homogenous transformation matrix describing the pose of an arbitrary frame with respect to the reference frame. Let us consider the following product of homogenous matrices which gives a new homogenous transformation matrix **H**

$$
\begin{aligned}
\mathbf{H} &= Trans(4,-3,7)Rot(y,90^\circ)Rot(z,90^\circ) \\
&= \begin{bmatrix} 1 & 0 & 0 & 4 \\ 0 & 1 & 0 & -3 \\ 0 & 0 & 1 & 7 \\ 0 & 0 & 0 & 1 \end{bmatrix} \begin{bmatrix} 0 & 0 & 1 & 0 \\ 0 & 1 & 0 & 0 \\ -1 & 0 & 0 & 0 \\ 0 & 0 & 0 & 1 \end{bmatrix} \begin{bmatrix} 0 & -1 & 0 & 0 \\ 1 & 0 & 0 & 0 \\ 0 & 0 & 1 & 0 \\ 0 & 0 & 0 & 1 \end{bmatrix} \\
&= \begin{bmatrix} 0 & 0 & 1 & 4 \\ 1 & 0 & 0 & -3 \\ 0 & 1 & 0 & 7 \\ 0 & 0 & 0 & 1 \end{bmatrix}.
\end{aligned}
\tag{2.11}
$$

When defining the homogenous matrix representing rotation, we learned that the first three columns describe the rotation of the frame x', y', z' with respect to the reference frame x, y, z

$$
\begin{array}{c}
\begin{matrix} x' & y' & z' & \end{matrix} \\
\begin{bmatrix} \begin{bmatrix} 0 \\ 1 \\ 0 \end{bmatrix} & \begin{bmatrix} 0 \\ 0 \\ 1 \end{bmatrix} & \begin{bmatrix} 1 \\ 0 \\ 0 \end{bmatrix} & \begin{matrix} 4 \\ -3 \\ 7 \end{matrix} \\ 0 & 0 & 0 & 1 \end{bmatrix} \begin{matrix} x \\ y \\ z \\ \end{matrix}
\end{array}.
\tag{2.12}
$$

The fourth column represents the position of the origin of the frame x', y', z' with respect to the reference frame x, y, z. With this knowledge we can represent graphically the frame x', y', z' described by the homogenous transformation matrix (2.11), relative to the reference frame x, y, z (Figure 2.5). The x' axis points in the direction of y axis of the reference frame, the y' axis is in the direction of the z axis, and the z' axis is in the x direction.

To convince ourselves of the correctness of the frame drawn in Figure 2.6, we shall check the displacements included in equation (2.11). The reference frame is

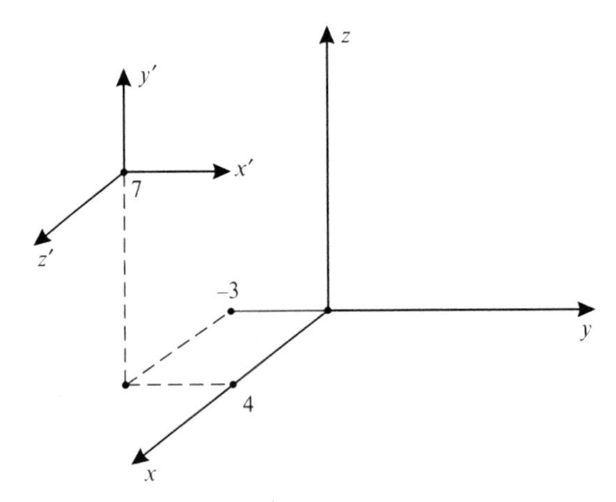

Fig. 2.5 The pose of an arbitrary frame $[x'\ y'\ z']$ with respect to the reference frame $[x\ y\ z]$

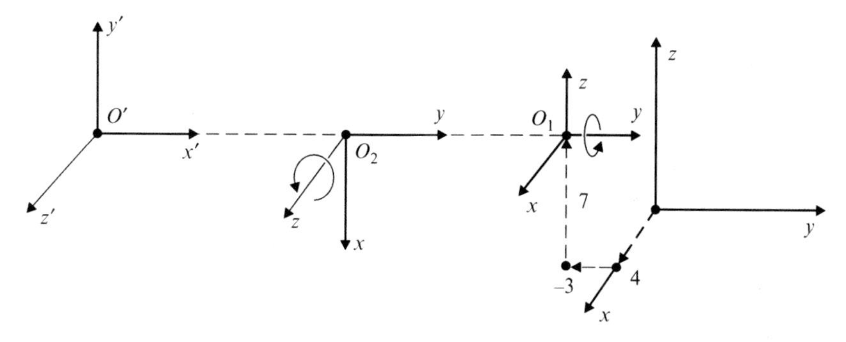

Fig. 2.6 Displacement of the reference frame into a new pose (from right to left). The origins O_1, O_2 and O' are in the same point

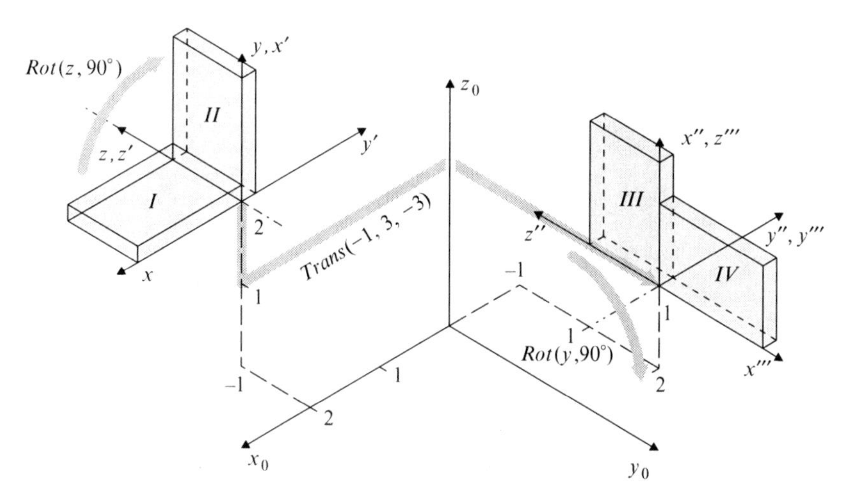

Fig. 2.7 Displacement of the object into a new pose

first translated into the point $[4, -3, 7]^T$, afterwards it is rotated for 90° around the new y axis and finally it is rotated for 90° around the newest z axis (Figure 2.6). The three displacements of the reference frame result in the same final pose as shown in Figure 2.5.

In continuation of this chapter we wish to elucidate the second meaning of the homogenous transformation matrix, i.e. a displacement of an object or coordinate frame into a new pose (Figure 2.7). First, we wish to rotate the coordinate frame x, y, z for 90° in the counter-clockwise direction around the z axis. This can be achieved by the following postmultiplication of the matrix $\mathbf{H}$ describing the initial pose of the coordinate frame x, y, z

$$\mathbf{H}_1 = \mathbf{H} \cdot Rot(z, 90^\circ). \tag{2.13}$$

The displacement resulted in a new pose of the object and new frame x', y', z' shown in Figure 2.7. We shall displace this new frame for -1 along the x' axis, 3 units along y' axis and -3 along z' axis

$$\mathbf{H}_2 = \mathbf{H}_1 \cdot Trans(-1,3,-3). \tag{2.14}$$

After translation a new pose of the object is obtained together with a new frame x'', y'', z''. This frame will be finally rotated for 90° around the y'' axis in the positive direction

$$\mathbf{H}_3 = \mathbf{H}_2 \cdot Rot(y'',90°). \tag{2.15}$$

The equations (2.13), (2.14) and (2.15) can be successively inserted one into another

$$\mathbf{H}_3 = \mathbf{H} \cdot Rot(z,90°) \cdot Trans(-1,3,-3) \cdot Rot(y'',90°) = \mathbf{H} \cdot \mathbf{D}. \tag{2.16}$$

In equation (2.16) the matrix $\mathbf{H}$ represents the initial pose of the frame, $\mathbf{H}_3$ is the final pose, while $\mathbf{D}$ represents the displacement

$$\begin{aligned}\mathbf{D} &= Rot(z,90°) \cdot Trans(-1,3,-3) \cdot Rot(y'',90°)\\ &= \begin{bmatrix} 0 & -1 & 0 & 0 \\ 1 & 0 & 0 & 0 \\ 0 & 0 & 1 & 0 \\ 0 & 0 & 0 & 1 \end{bmatrix} \begin{bmatrix} 1 & 0 & 0 & -1 \\ 0 & 1 & 0 & 3 \\ 0 & 0 & 1 & -3 \\ 0 & 0 & 0 & 1 \end{bmatrix} \begin{bmatrix} 0 & 0 & 1 & 0 \\ 0 & 1 & 0 & 0 \\ -1 & 0 & 0 & 0 \\ 0 & 0 & 0 & 1 \end{bmatrix}\\ &= \begin{bmatrix} 0 & -1 & 0 & -3 \\ 0 & 0 & 1 & -1 \\ -1 & 0 & 0 & -3 \\ 0 & 0 & 0 & 1 \end{bmatrix}.\end{aligned} \tag{2.17}$$

Finally we shall perform the postmultiplication describing the new relative pose of the object

$$\begin{aligned}\mathbf{H}_3 = \mathbf{H} \cdot \mathbf{D} &= \begin{bmatrix} 1 & 0 & 0 & 2 \\ 0 & 0 & -1 & -1 \\ 0 & 1 & 0 & 2 \\ 0 & 0 & 0 & 1 \end{bmatrix} \begin{bmatrix} 0 & -1 & 0 & -3 \\ 0 & 0 & 1 & -1 \\ -1 & 0 & 0 & -3 \\ 0 & 0 & 0 & 1 \end{bmatrix}\\ &= \begin{array}{c} \begin{matrix} x''' & y''' & z''' & \end{matrix} \\ \begin{bmatrix} 0 & -1 & 0 & -1 \\ 1 & 0 & 0 & 2 \\ 0 & 0 & 1 & 1 \\ 0 & 0 & 0 & 1 \end{bmatrix} \begin{matrix} x_0 \\ y_0 \\ z_0 \\ \end{matrix} \end{array}.\end{aligned} \tag{2.18}$$

As in the previous example we shall graphically verify the correctness of the matrix (2.18). The three displacements of the frame x, y, z: rotation for 90° in counter-clockwise direction around the z axis, translation for -1 along the x' axis,

3 units along y' axis and -3 along z' axis, and rotation for 90° around y'' axis in the positive direction are shown in Figure 2.7. The result is the final pose of the object x''', y''', z'''. The x''' axis points in the positive direction of the y_0 axis, y''' points in the negative direction of x_0 axis and z''' points in the positive direction of z_0 axis of the reference frame. The directions of the axes of the final frame correspond to the first three columns of the matrix $\mathbf{H}_3$. There is also agreement between the position of the origin of the final frame in Figure 2.7 and the fourth column of the matrix $\mathbf{H}_3$.

2.4 Geometrical robot model

Our final goal is the geometrical model of a robot manipulator. A geometrical robot model is given by the description of the pose of the last segment of the robot (end-effector) expressed in the reference (base) frame. The knowledge how to describe the pose of an object by the use of homogenous transformation matrices will be first applied to the process of assembly. For this purpose a mechanical assembly consisting of four blocks, such as presented in Figure 2.8, will be considered. A plate with dimensions $(5 \times 15 \times 1)$ is placed over a block $(5 \times 4 \times 10)$. Another plate $(8 \times 4 \times 1)$ is positioned perpendicularly to the first one, holding another small block $(1 \times 1 \times 5)$.

A frame is attached to each of the four blocks as shown in Figure 2.8. Our task will be to calculate the pose of the O_3 frame with respect to the reference frame O_0. In the last chapter we learned that the pose of a displaced frame can be expressed

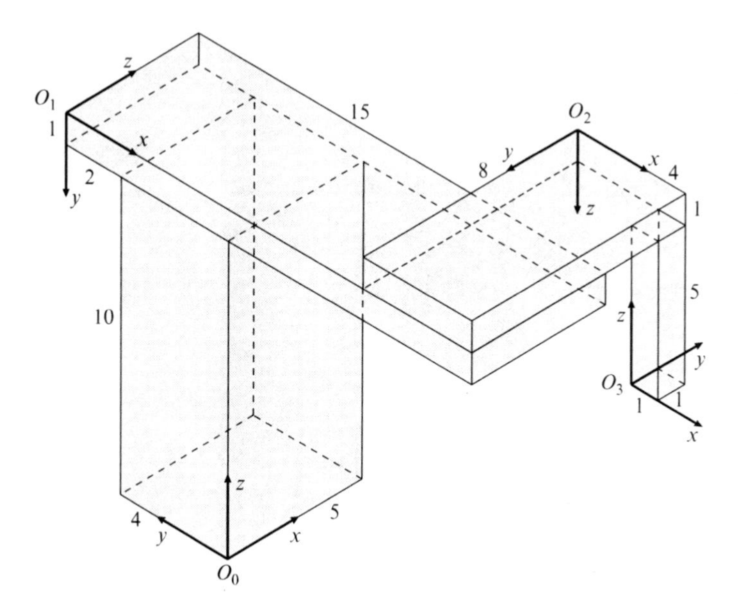

Fig. 2.8 Mechanical assembly

with respect to the reference frame by the use of the homogenous transformation matrix $\mathbf{H}$. The pose of the frame O_1 with respect to the frame O_0 will be denoted by ${}^0\mathbf{H}_1$. In the same way ${}^1\mathbf{H}_2$ represents the pose of O_2 frame with respect to O_1 and ${}^2\mathbf{H}_3$ the pose of O_3 with regard to O_2 frame. We learned also that the successive displacements are expressed by postmultiplications (successive multiplications from left to right) of homogenous transformation matrices. Also the assembly process can be described by postmultiplication of the corresponding matrices. The pose of the fourth block can be written with respect to the first one by the following matrix

$$ {}^0\mathbf{H}_3 = {}^0\mathbf{H}_1 {}^1\mathbf{H}_2 {}^2\mathbf{H}_3. \tag{2.19} $$

The blocks were positioned perpendicularly one to another. In this way it is not necessary to calculate the sines and cosines of the rotation angles. The matrices can be determined directly from Figure 2.8. The x axis of O_1 frame points in negative direction of the y axis in the O_0 frame. The y axis of O_1 frame points in negative direction of the z axis in the O_0 frame. The z axis of the O_1 frame has the same direction as x axis of the O_0 frame. The described geometrical properties of the assembly structure are written into the first three columns of the homogenous matrix. The position of the origin of the O_1 frame with respect to the O_0 frame is written into the fourth column

$$ {}^0\mathbf{H}_1 = \begin{array}{c} \overbrace{\begin{array}{ccc} x & y & z \end{array}}^{O_1} \\ \left[\begin{array}{cccc} 0 & 0 & 1 & 0 \\ -1 & 0 & 0 & 6 \\ 0 & -1 & 0 & 11 \\ 0 & 0 & 0 & 1 \end{array}\right] \end{array} \left.\begin{array}{c} x \\ y \\ z \end{array}\right\} O_0 . \tag{2.20} $$

In the same way the other two matrices are determined

$$ {}^1\mathbf{H}_2 = \begin{bmatrix} 1 & 0 & 0 & 11 \\ 0 & 0 & 1 & -1 \\ 0 & -1 & 0 & 8 \\ 0 & 0 & 0 & 1 \end{bmatrix} \tag{2.21} $$

$$ {}^2\mathbf{H}_3 = \begin{bmatrix} 1 & 0 & 0 & 3 \\ 0 & -1 & 0 & 1 \\ 0 & 0 & -1 & 6 \\ 0 & 0 & 0 & 1 \end{bmatrix} . \tag{2.22} $$

The position and orientation of the fourth block with respect to the first one is given by the ${}^0\mathbf{H}_3$ matrix which is obtained by successive multiplication of the matrices (2.20), (2.21) and (2.22)

$$ {}^0\mathbf{H}_3 = \begin{bmatrix} 0 & 1 & 0 & 7 \\ -1 & 0 & 0 & -8 \\ 0 & 0 & 1 & 6 \\ 0 & 0 & 0 & 1 \end{bmatrix}. \tag{2.23} $$

The fourth column of the matrix ${}^0\mathbf{H}_3[7,-8,6,1]^T$ represents the position of the origin of the O_3 frame with respect to the reference frame O_0. The correctness of the fourth column can be checked from Figure 2.8. The rotational part of the matrix ${}^0\mathbf{H}_3$ represents the orientation of the O_3 frame with respect to the reference frame O_0.

Now let us imagine that the first horizontal plate rotates with respect to the first vertical block around axis 1 for angle ϑ_1. The second plate also rotates around the vertical axis 2 for angle ϑ_2. The last block is elongated for distance d_3 along the third axis. In this way we obtained a robot manipulator, which was named SCARA in the introductory chapter.

Our goal is to develop a geometrical model of the SCARA robot. Blocks and plates from Figure 2.9 will be replaced by symbols for rotational and translational joints that we know from the introduction (Figure 2.10).

The first vertical segment with the length l_1 starts from the basis, where the robot is attached to the ground, and is terminated by the first rotational joint. The second segment with length l_2 is horizontal and rotates around the first segment. The rotation in the first joint is denoted by the angle ϑ_1. The third segment with the length l_3 is also horizontal and rotates around the vertical axis at the end of the second

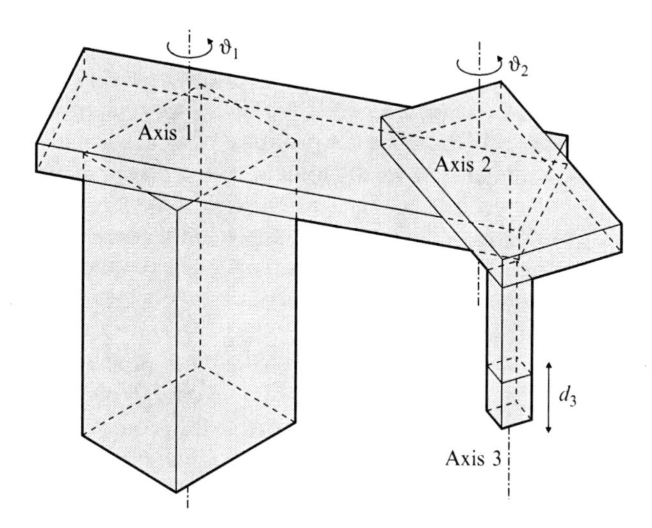

Fig. 2.9 Displacements of the mechanical assembly

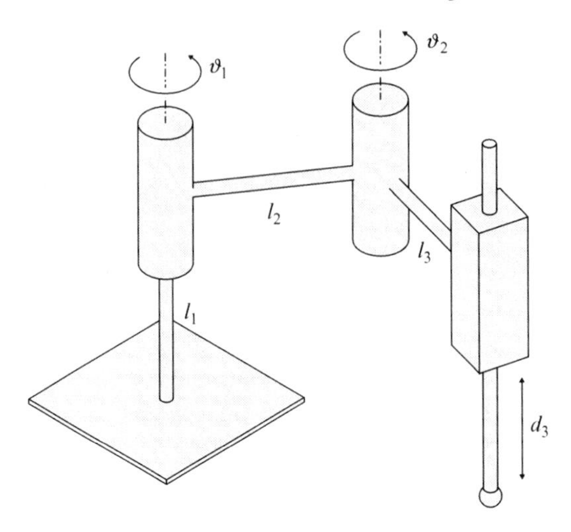

Fig. 2.10 SCARA robot manipulator in an arbitrary pose

segment. The angle is denoted as ϑ_2. There is a translational joint at the end of the third segment. It enables the robot end-effector to approach the working plane where the robot task takes place. The translational joint is displaced from zero initial length to the length described by the variable d_3.

The robot mechanism is first brought to the initial pose which is also called "home position". In the initial pose two neighboring segments must be either parallel or perpendicular. The translational joints are in their initial position $d_i = 0$. The initial pose of the SCARA manipulator is shown in Figure 2.11.

First, the coordinate frames must be drawn into the SCARA robot presented in Figure 2.11. The first (reference) coordinate frame x_0, y_0, z_0 is placed onto the base of the robot. In the last chapter we shall learn that robot standards require the z_0 axis to point perpendicularly out from the base. In this case it is aligned with the first segment. The other two axes are selected in such a way that robot segments are parallel to one of the axes of the reference coordinate frame, when the robot is in its initial home position. In our case we align the y_0 axis with the segments l_2 and l_3. The coordinate frame must be right handed. The rest of the frames are placed into the robot joints. The origins of the frames are drawn in the center of each joint. One of the frame axes must be aligned with the joint axis. The simplest way to calculate the geometrical model of a robot is to make all the frames in the robot joints parallel to the reference frame (Figure 2.11).

The geometrical model of a robot describes the pose of the frame attached to the end-effector with respect to the reference frame on the robot base. Similar to the case of the mechanical assembly, we shall obtain the geometrical model by successive multiplication (postmultiplication) of homogenous transformation matrices. The main difference between the mechanical assembly and the robot manipulator are the displacements of robot joints. For this purpose, each matrix ${}^{i-1}\mathbf{H}_i$ describing the pose of a segment will be followed by a matrix $\mathbf{D}_i$ representing the displacement

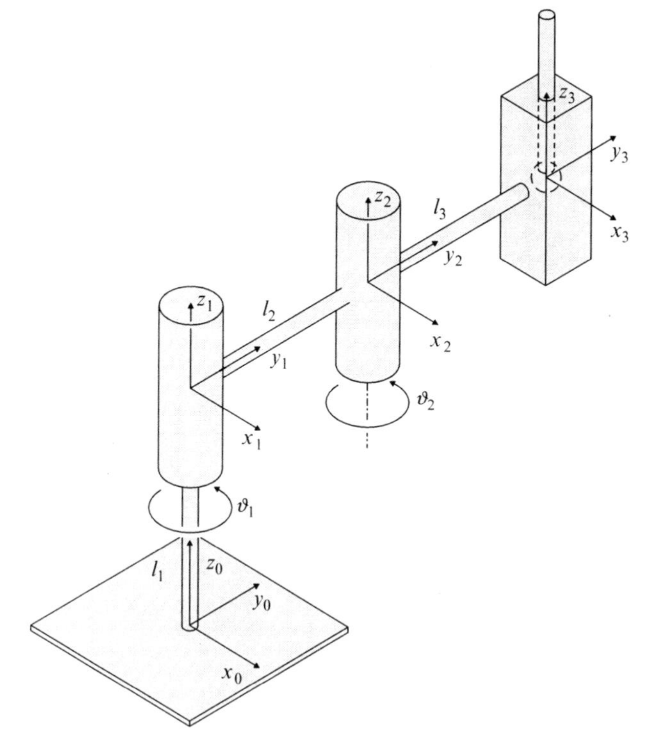

Fig. 2.11 The SCARA robot manipulator in the initial pose

of either the translational or the rotational joint. Our SCARA robot has three joints. The pose of the end frame x_3, y_3, z_3 with respect to the base frame x_0, y_0, z_0 is expressed by the following postmultiplication of three pairs of homogenous transformation matrices

$$^{0}\mathbf{H}_3 = (^{0}\mathbf{H}_1\mathbf{D}_1)\cdot(^{1}\mathbf{H}_2\mathbf{D}_2)\cdot(^{2}\mathbf{H}_3\mathbf{D}_3). \tag{2.24}$$

In equation (2.24) the matrices $^{0}\mathbf{H}_1$, $^{1}\mathbf{H}_2$, and $^{2}\mathbf{H}_3$ describe the pose of each joint frame with respect to the preceding frame in the same way as in the case of assembly of the blocs. From Figure 2.11 it is evident that the $\mathbf{D}_1$ matrix represents a rotation around the positive z_1 axis. The following product of two matrices describes the pose and the displacement in the first joint

$$^{0}\mathbf{H}_1\mathbf{D}_1 = \begin{bmatrix} 1 & 0 & 0 & 0 \\ 0 & 1 & 0 & 0 \\ 0 & 0 & 1 & l_1 \\ 0 & 0 & 0 & 1 \end{bmatrix} \begin{bmatrix} c1 & -s1 & 0 & 0 \\ s1 & c1 & 0 & 0 \\ 0 & 0 & 1 & 0 \\ 0 & 0 & 0 & 1 \end{bmatrix} = \begin{bmatrix} c1 & -s1 & 0 & 0 \\ s1 & c1 & 0 & 0 \\ 0 & 0 & 1 & l_1 \\ 0 & 0 & 0 & 1 \end{bmatrix}.$$

In the above matrices the following shorter notation was used: $\sin\vartheta_1 = s1$ and $\cos\vartheta_1 = c1$.

In the second joint there is a rotation around the z_2 axis

$$^{1}\mathbf{H}_2\mathbf{D}_2 = \begin{bmatrix} 1 & 0 & 0 & 0 \\ 0 & 1 & 0 & l_2 \\ 0 & 0 & 1 & 0 \\ 0 & 0 & 0 & 1 \end{bmatrix} \begin{bmatrix} c2 & -s2 & 0 & 0 \\ s2 & c2 & 0 & 0 \\ 0 & 0 & 1 & 0 \\ 0 & 0 & 0 & 1 \end{bmatrix} = \begin{bmatrix} c2 & -s2 & 0 & 0 \\ s2 & c2 & 0 & l_2 \\ 0 & 0 & 1 & 0 \\ 0 & 0 & 0 & 1 \end{bmatrix}.$$

In the last joint there is translation along the z_3 axis

$$^{2}\mathbf{H}_3\mathbf{D}_3 = \begin{bmatrix} 1 & 0 & 0 & 0 \\ 0 & 1 & 0 & l_3 \\ 0 & 0 & 1 & 0 \\ 0 & 0 & 0 & 1 \end{bmatrix} \begin{bmatrix} 1 & 0 & 0 & 0 \\ 0 & 1 & 0 & 0 \\ 0 & 0 & 1 & -d_3 \\ 0 & 0 & 0 & 1 \end{bmatrix} = \begin{bmatrix} 1 & 0 & 0 & 0 \\ 0 & 1 & 0 & l_3 \\ 0 & 0 & 1 & -d_3 \\ 0 & 0 & 0 & 1 \end{bmatrix}.$$

The geometrical model of the SCARA robot manipulator is obtained by postmultiplication of the three matrices derived above

$$^{0}\mathbf{H}_3 = \begin{bmatrix} c12 & -s12 & 0 & -l_3 s12 - l_2 s1 \\ s12 & c12 & 0 & l_3 c12 + l_2 c1 \\ 0 & 0 & 1 & l_1 - d_3 \\ 0 & 0 & 0 & 1 \end{bmatrix}.$$

When multiplying the three matrices the following abbreviation was introduced $c12 = \cos(\vartheta_1 + \vartheta_2) = c1c2 - s1s2$ and $s12 = \sin(\vartheta_1 + \vartheta_2) = s1c2 + c1s2$.

Chapter 3
Geometric description of the robot mechanism

The geometric description of the robot mechanism is based on the usage of translational and rotational homogenous transformation matrices. A coordinate frame is attached to the robot base and to each segment of the mechanism, as shown in Figure 3.1. Then, the corresponding transformation matrices between the consecutive frames are determined. A vector expressed in one of the frames can be transformed into another frame by successive multiplication of intermediate transformation matrices.

Vector **a** in Figure 3.1 is expressed relative to the coordinate frame x_3, y_3, z_3, while vector **b** is given in the frame x_0, y_0, z_0 belonging to the robot base. A mathematical relation between the two vectors is obtained by the following homogenous transformation

$$\begin{bmatrix} \mathbf{b} \\ 1 \end{bmatrix} = {}^0\mathbf{H}_1 {}^1\mathbf{H}_2 {}^2\mathbf{H}_3 \begin{bmatrix} \mathbf{a} \\ 1 \end{bmatrix}. \tag{3.1}$$

3.1 Vector parameters of a kinematic pair

Vector parameters will be used for the geometric description of a robot mechanism. For simplicity we shall limit our consideration to the mechanisms with either parallel or perpendicular consecutive joint axes. Such mechanisms are by far the most frequent in industrial robotics.

In Figure 3.2, a kinematic pair is shown consisting of two consecutive segments of a robot mechanism, segment $i-1$ and segment i. The two segments are connected by the joint i including both translation and rotation. The relative pose of the joint is determined by the segment vector $\mathbf{b}_{i-1}$ and unit joint vector $\mathbf{e}_i$, as shown in the Figure 3.2. The segment i can be translated with respect to the segment $i-1$ along the vector $\mathbf{e}_i$ for the distance d_i and can be rotated around $\mathbf{e}_i$ for the angle ϑ_i. The coordinate frame x_i, y_i, z_i is attached to the segment i, while the frame x_{i-1}, y_{i-1}, z_{i-1} belongs to the segment $i-1$.

T. Bajd et al., *Robotics*, Intelligent Systems, Control and Automation: Science and Engineering 43, DOI 10.1007/978-90-481-3776-3_3,

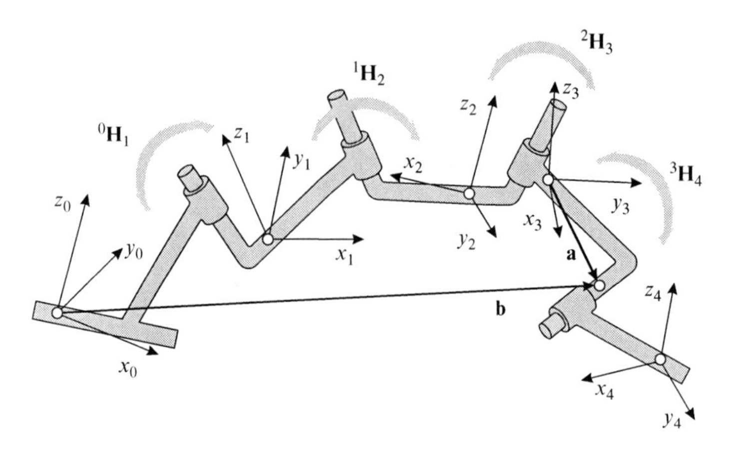

Fig. 3.1 Robot mechanism with coordinate frames attached to its segments

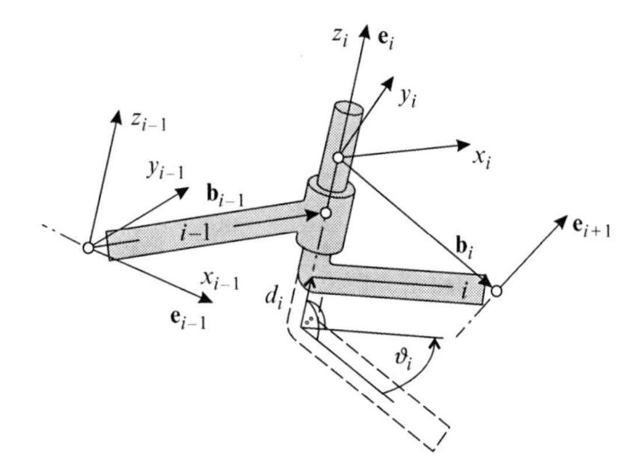

Fig. 3.2 Vector parameters of a kinematic pair

The coordinate frame x_i, y_i, z_i is placed into the axis of the joint i in such a way that it is parallel to the previous frame x_{i-1}, y_{i-1}, z_{i-1} when the kinematic pair is in its initial pose (both joint variables are zero $\vartheta_i = 0$ and $d_i = 0$).

The geometric relations and the relative displacement of two neighboring segments of a robot mechanism are determined by the following parameters:

$\mathbf{e}_i$ – unit vector describing either the axis of rotation or direction of translation in the joint i and is expressed as one of the axes of the x_i, y_i, z_i frame. Its components are the following

$$\mathbf{e}_i = \begin{bmatrix} 1 \\ 0 \\ 0 \end{bmatrix} \text{ or } \begin{bmatrix} 0 \\ 1 \\ 0 \end{bmatrix} \text{ or } \begin{bmatrix} 0 \\ 0 \\ 1 \end{bmatrix};$$

$\mathbf{b}_{i-1}$ – segment vector describing the segment $i-1$ expressed in the x_{i-1}, y_{i-1}, z_{i-1} frame. Its components are the following

$$\mathbf{b}_{i-1} = \begin{bmatrix} b_{i-1,x} \\ b_{i-1,y} \\ b_{i-1,z} \end{bmatrix};$$

ϑ_i – rotational variable representing the angle measured around the $\mathbf{e}_i$ axis in the plane which is perpendicular to $\mathbf{e}_i$ (the angle is zero when the kinematic pair is in the initial position)

d_i – translational variable representing the distance measured along the direction of $\mathbf{e}_i$ (the distance equals zero when the kinematic pair is in the initial position)

If the joint is only rotational (Figure 3.3), the joint variable is represented by the angle ϑ_i, while $d_i = 0$. When the robot mechanism is in its initial pose, the joint angle equals zero $\vartheta_i = 0$ and the coordinate frames x_i, y_i, z_i and x_{i-1}, y_{i-1}, z_{i-1} are parallel. If the joint is only translational (Figure 3.3), the joint variable is d_i, while $\vartheta_i = 0$. When the joint is in its initial position, then $d_i = 0$. In this case the coordinate frames x_i, y_i, z_i and x_{i-1}, y_{i-1}, z_{i-1} are parallel irrespective of the value of the translational variable d_i.

By changing the value of the rotational joint variable ϑ_i, the coordinate frame x_i, y_i, z_i is rotated together with the segment i with respect to the preceding segment $i-1$ and the corresponding x_{i-1}, y_{i-1}, z_{i-1} frame. By changing the translational variable d_i, the displacement is translational, where only the distance between the two neighboring frames is changing.

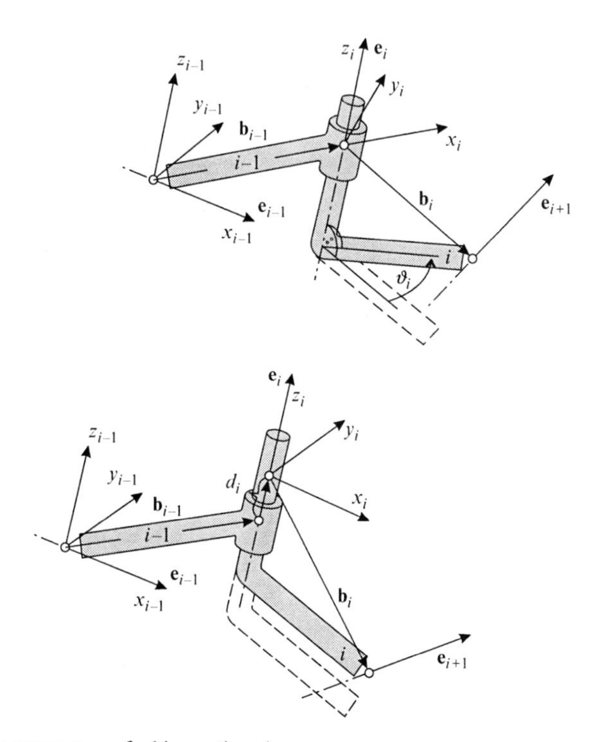

Fig. 3.3 Vector parameters of a kinematic pair

The transformation between the coordinate frames x_{i-1}, y_{i-1}, z_{i-1} and x_i, y_i, z_i is determined by the homogenous transformation matrix taking one of the three possible forms regarding the direction of the joint vector $\mathbf{e}_i$. When the unit vector $\mathbf{e}_i$ is parallel to the x_i axis, there is

$$ {}^{i-1}\mathbf{H}_i = \begin{bmatrix} 1 & 0 & 0 & d_i + b_{i-1,x} \\ 0 & \cos\vartheta_i & -\sin\vartheta_i & b_{i-1,y} \\ 0 & \sin\vartheta_i & \cos\vartheta_i & b_{i-1,z} \\ 0 & 0 & 0 & 1 \end{bmatrix}, \tag{3.2} $$

when $\mathbf{e}_i$ is parallel to the y_i axis, we have the following transformation matrix

$$ {}^{i-1}\mathbf{H}_i = \begin{bmatrix} \cos\vartheta_i & 0 & \sin\vartheta_i & b_{i-1,x} \\ 0 & 1 & 0 & d_i + b_{i-1,y} \\ -\sin\vartheta_i & 0 & \cos\vartheta_i & b_{i-1,z} \\ 0 & 0 & 0 & 1 \end{bmatrix} \tag{3.3} $$

and when $\mathbf{e}_i$ is parallel to the z_i axis, the matrix has the following form

$$ {}^{i-1}\mathbf{H}_i = \begin{bmatrix} \cos\vartheta_i & -\sin\vartheta_i & 0 & b_{i-1,x} \\ \sin\vartheta_i & \cos\vartheta_i & 0 & b_{i-1,y} \\ 0 & 0 & 1 & d_i + b_{i-1,z} \\ 0 & 0 & 0 & 1 \end{bmatrix}. \tag{3.4} $$

In the initial pose the coordinate frames x_{i-1}, y_{i-1}, z_{i-1} and x_i, y_i, z_i are parallel ($\vartheta_i = 0$ and $d_i = 0$) and displaced only for the vector $\mathbf{b}_{i-1}$

$$ {}^{i-1}\mathbf{H}_i = \begin{bmatrix} 1 & 0 & 0 & b_{i-1,x} \\ 0 & 1 & 0 & b_{i-1,y} \\ 0 & 0 & 1 & b_{i-1,z} \\ 0 & 0 & 0 & 1 \end{bmatrix}. \tag{3.5} $$

3.2 Vector parameters of the mechanism

The vector parameters of a robot mechanism are determined in the following four steps:

Step 1 – The robot mechanism is placed into the desired initial (reference) pose. The joint axes must be parallel to one of the axes of the reference coordinate frame x_0, y_0, z_0 attached to the robot base. In the reference pose all values of joint variables equal zero, $\vartheta_i = 0$ and $d_i = 0$, $i = 1, 2, ..., n$

Step 2 – The centers of the joints $i = 1, 2, ..., n$ are selected. The center of joint i can be anywhere along the corresponding joint axis. A local coordinate frame x_i, y_i, z_i is placed into the joint center in such a way that its axes are parallel to the axes of the reference frame x_0, y_0, z_0. The local coordinate frame x_i, y_i, z_i is displaced together with the segment i

Step 3 – The unit joint vector $\mathbf{e}_i$ is allocated to each joint axis $i = 1, 2, \ldots, n$. It is directed along one of the axes of the coordinate frame x_i, y_i, z_i. In the direction of this vector the translational variable d_i is measured, while the rotational variable ϑ_i is assessed around the joint vector $\mathbf{e}_i$

Step 4 – The segment vectors $\mathbf{b}_{i-1}$ are drawn between the origins of the x_i, y_i, z_i frames, $i = 1, 2, \ldots, n$. The segment vector $\mathbf{b}_n$ connects the origin of the x_n, y_n, z_n frame with the robot end-point

Sometimes an additional coordinate frame is positioned in the reference point of a gripper and denoted as x_{n+1}, y_{n+1}, z_{n+1}. There exists no degree of freedom between the frames x_n, y_n, z_n and x_{n+1}, y_{n+1}, z_{n+1}, as both frames are attached to the same segment. The transformation between them is therefore constant.

The approach to geometric modeling of robot mechanisms will be illustrated by an example of a robot mechanism with four degrees of freedom shown in Figure 3.4. The selected initial pose of the mechanism together with the marked positions of the joint centers is presented in Figure 3.5. The corresponding vector parameters and joint variables are gathered in Table 3.1.

The rotational variables ϑ_1, ϑ_2 and ϑ_4 are measured in the planes perpendicular to the joint axes $\mathbf{e}_1$, $\mathbf{e}_2$ and $\mathbf{e}_4$, while the translational variable d_i is measured along

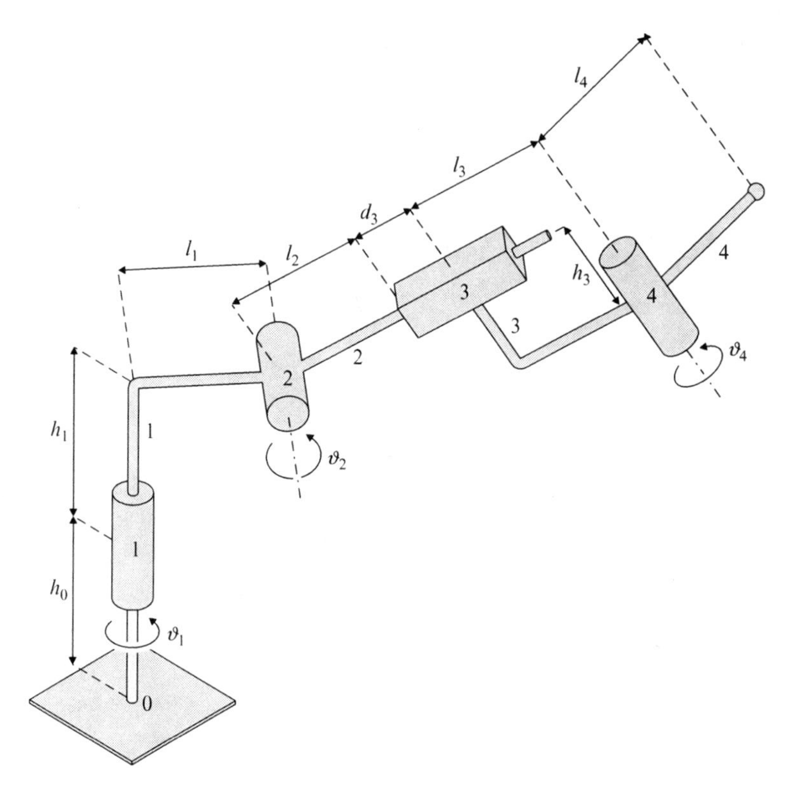

Fig. 3.4 Robot mechanism with four degrees of freedom

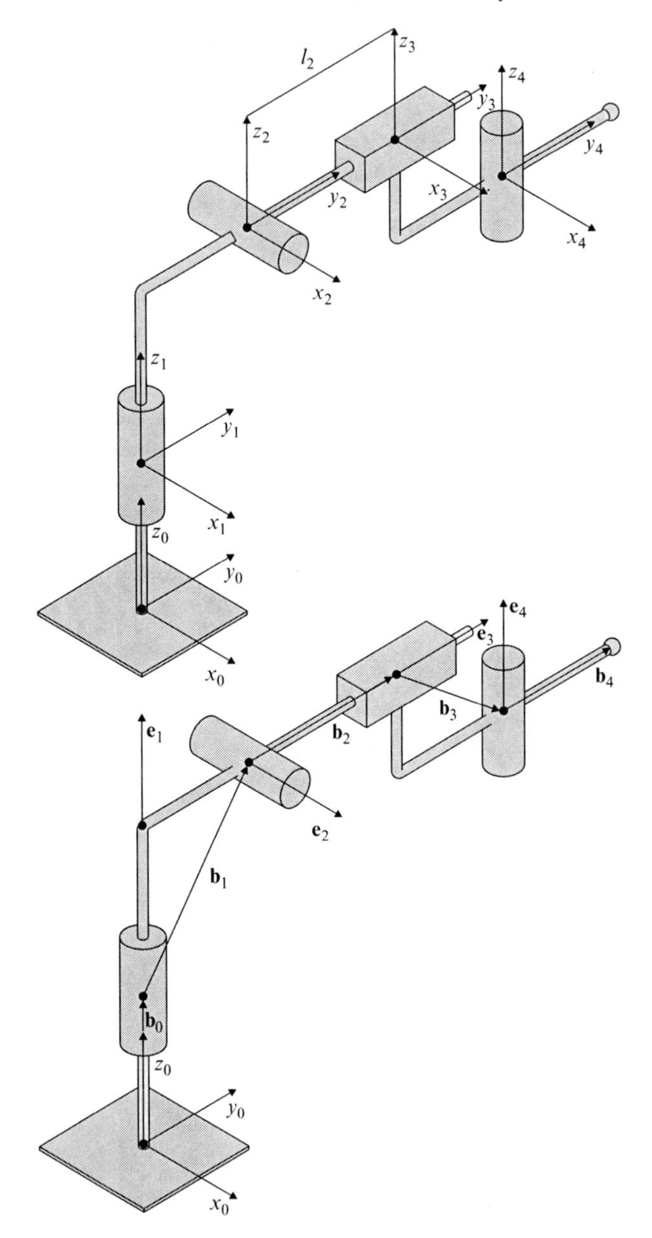

Fig. 3.5 Positioning of the coordinate frames for the robot mechanism with four degrees of freedom

the axis $\mathbf{e}_3$. Their values are zero when the robot mechanism is in its initial pose. In Figure 3.6 the robot manipulator is shown in a pose where all four variables are positive and nonzero. The variable ϑ_1 represents the angle between the initial and

Table 3.1 Vector parameters and joint variables for the robot mechanism in Figure 3.5

i	1	2	3	4
ϑ_i	ϑ_1	ϑ_2	0	ϑ_4
d_i	0	0	d_3	0

i	1	2	3	4
	0	1	0	0
$\mathbf{e}_i$	0	0	1	0
	1	0	0	1

i	1	2	3	4	5
	0	0	0	0	0
$\mathbf{b}_{i-1}$	0	l_1	l_2	l_3	l_4
	h_0	h_1	0	$-h_3$	0

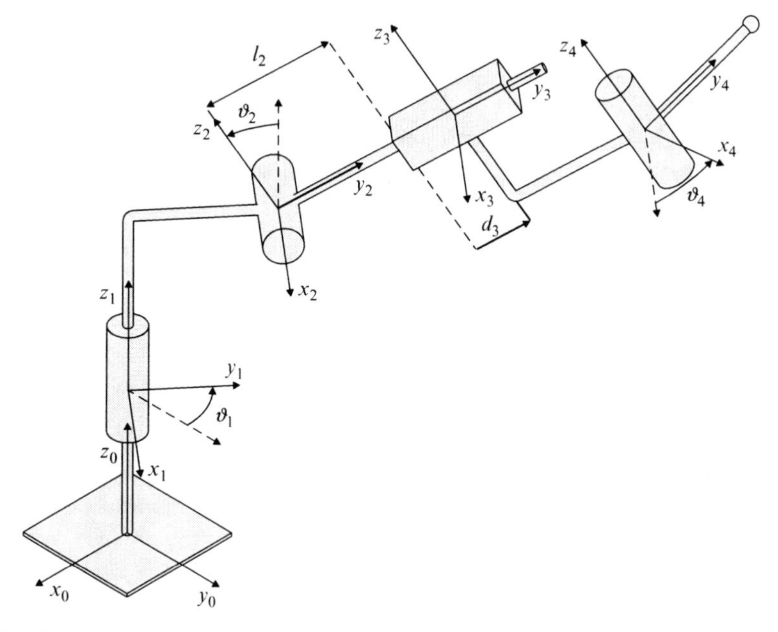

Fig. 3.6 Determining the rotational and translational variables for the robot mechanism with four degrees of freedom

momentary y_1 axis, the variable ϑ_2 the angle between the initial and momentary z_2 axis, variable d_3 is the distance between the initial and actual position of the x_3 axis, while ϑ_4 represents the angle between the initial and momentary x_4 axis.

The selected vector parameters of the robot mechanism are inserted into the homogenous transformation matrices (3.2)–(3.4)

$$
{}^0\mathbf{H}_1 = \begin{bmatrix} c1 & -s1 & 0 & 0 \\ s1 & c1 & 0 & 0 \\ 0 & 0 & 1 & h_0 \\ 0 & 0 & 0 & 1 \end{bmatrix},
$$

$$
{}^1\mathbf{H}_2 = \begin{bmatrix} 1 & 0 & 0 & 0 \\ 0 & c2 & -s2 & l_1 \\ 0 & s2 & c2 & h_1 \\ 0 & 0 & 0 & 1 \end{bmatrix},
$$

$$
{}^2\mathbf{H}_3 = \begin{bmatrix} 1 & 0 & 0 & 0 \\ 0 & 1 & 0 & d_3 + l_2 \\ 0 & 0 & 1 & 0 \\ 0 & 0 & 0 & 1 \end{bmatrix},
$$

$$
{}^3\mathbf{H}_4 = \begin{bmatrix} c4 & -s4 & 0 & 0 \\ s4 & c4 & 0 & l_3 \\ 0 & 0 & 1 & -h_3 \\ 0 & 0 & 0 & 1 \end{bmatrix}.
$$

An additional homogenous matrix describes the position of the gripper reference point where the coordinate frame x_5, y_5, z_5 can be allocated

$$
{}^4\mathbf{H}_5 = \begin{bmatrix} 1 & 0 & 0 & 0 \\ 0 & 1 & 0 & l_4 \\ 0 & 0 & 1 & 0 \\ 0 & 0 & 0 & 1 \end{bmatrix}.
$$

This last matrix is constant as the frames x_4, y_4, z_4 and x_5, y_5, z_5 are parallel and displaced for the distance l_4. Usually this additional frame is not even attached to the robot mechanism, as the position and orientation of the gripper can be described in the x_4, y_4, z_4 frame.

When determining the initial (home) pose of the robot mechanism we must take care that the joint axes are parallel to one of the axes of the reference coordinate frame. The initial pose should be selected in such a way that it is simple and easy to examine, that it corresponds well to the anticipated robot tasks and that it minimizes the number of required mathematical operations included in the transformation matrices.

As another example we shall consider the SCARA robot manipulator whose geometric model was developed already in the previous chapter and is shown in Figure 2.10. The robot mechanism should be first positioned into the initial pose

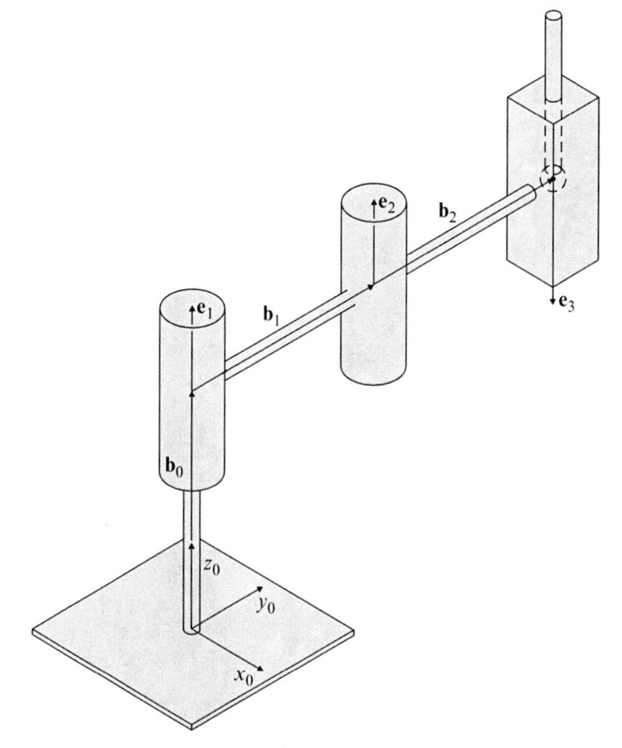

Fig. 3.7 The SCARA robot manipulator in the initial pose

in such a way that the joint axes are parallel to one of the axes of the reference frame x_0, y_0, z_0. In this way the two neighboring segments are either parallel or perpendicular. The translational joint must be in its initial position $(d3 = 0)$. The SCARA robot in the selected initial pose is shown in Figure 3.7.

The joint coordinate frames x_i, y_i, z_i are all parallel to the reference frame. Therefore, we shall draw only the reference frame and have the dots indicate the joint centers. In the centers of both rotational joints, unit vectors $\mathbf{e_1}$ and $\mathbf{e_2}$ are placed along the joint axes. The rotation around the $\mathbf{e_1}$ vector is described by the variable ϑ_1, while ϑ_2 represents the angle about the $\mathbf{e_2}$ vector. Vector $\mathbf{e_3}$ is placed along the translational axis of the third joint. Its translation variable is described by d_3. The first joint is connected to the robot base by the vector $\mathbf{b_0}$. Vector $\mathbf{b_1}$ connects the first and the second joint and vector $\mathbf{b_2}$ the second and the third joint. The variables and vectors are gathered in the three tables (Table 3.2).

In our case all $\mathbf{e_i}$ vectors are parallel to the z_0 axis, the homogenous transformation matrices are therefore written according equation (3.4). Similar matrices are obtained for both rotational joints.

Table 3.2 Vector parameters and joint variables for the SCARA robot manipulator

i	1	2	3
ϑ_i	ϑ_1	ϑ_2	0
d_i	0	0	d_3

i	1	2	3
$\mathbf{e}_i$	0 0 1	0 0 1	0 0 -1

i	1	2	3
$\mathbf{b}_{i-1}$	0 0 l_1	0 l_2 0	0 l_3 0

$$
{}^0\mathbf{H}_1 = \begin{bmatrix} c1 & -s1 & 0 & 0 \\ s1 & c1 & 0 & 0 \\ 0 & 0 & 1 & l_1 \\ 0 & 0 & 0 & 1 \end{bmatrix}.
$$

$$
{}^1\mathbf{H}_2 = \begin{bmatrix} c2 & -s2 & 0 & 0 \\ s2 & c2 & 0 & l_2 \\ 0 & 0 & 1 & 0 \\ 0 & 0 & 0 & 1 \end{bmatrix}.
$$

For the translational joint, $\vartheta_3 = 0$ must be inserted into equation (3.4), giving

$$
{}^2\mathbf{H}_3 = \begin{bmatrix} 1 & 0 & 0 & 0 \\ 0 & 1 & 0 & l_3 \\ 0 & 0 & 1 & -d_3 \\ 0 & 0 & 0 & 1 \end{bmatrix}.
$$

With postmultiplication of all three matrices the geometric model of the SCARA robot is obtained

$$
{}^0\mathbf{H}_3 = {}^0\mathbf{H}_1 {}^1\mathbf{H}_2 {}^2\mathbf{H}_3 = \begin{bmatrix} c12 & -s12 & 0 & -l_3 s12 - l_2 s1 \\ s12 & c12 & 0 & l_3 c12 + l_2 c1 \\ 0 & 0 & 1 & l_1 - d_3 \\ 0 & 0 & 0 & 1 \end{bmatrix}.
$$

We obtained the same result as in previous chapter, however in a much simpler and more clear way.

Chapter 4
Two-segment robot manipulator

4.1 Kinematics

Kinematics is part of mechanics studying motion without considering the forces which are responsible for this motion. Motion is in general described by trajectories, velocities and accelerations. In robotics we are mainly interested in trajectories and velocities, as both can be measured by the joint sensors. In the robot joints, the trajectories are measured either as the angle in a rotational joint or as the distance in a translational joint. The joint variables are also called internal coordinates. When planning and programming a robot task the trajectory of the robot end-point is of utmost importance. Position and orientation of the end-effector is described by external coordinates. Computation of external variables from the internal and vice versa is the central problem of robot kinematics.

In this chapter we shall limit our interest to a planar two-segment robot manipulator with two rotational joints (Figure 4.1). According to the definition from the introductory chapter, such a mechanism cannot even be called a robot. Nevertheless, this mechanism is an important part of the SCARA and anthropomorphic robot structures and will enable us to study several characteristic properties of the motion of robot mechanisms.

We distinguish between direct and inverse kinematics. Direct kinematics in the case of a two-segment robot represents the calculation of the position of the robot end-point from the known joint angles. Inverse kinematics calculates the joint variables from the known position of the robot end-point. Direct kinematics represents the simpler problem, as we have a single solution for the position of the robot end-point. The solutions of inverse kinematics depend largely on the structure of the robot manipulator. We often deal with several solutions for the joint variables resulting in the same position of the robot end-point, while in some cases an analytic solution of inverse kinematics does not exist.

Kinematic analysis includes also the relations between the velocity of the robot end-point and the velocities of individual joints. We shall find out that inverse kinematics for the velocities is simpler than inverse kinematics for the

T. Bajd et al., *Robotics*, Intelligent Systems, Control and Automation: Science and Engineering 43, DOI 10.1007/978-90-481-3776-3_4,

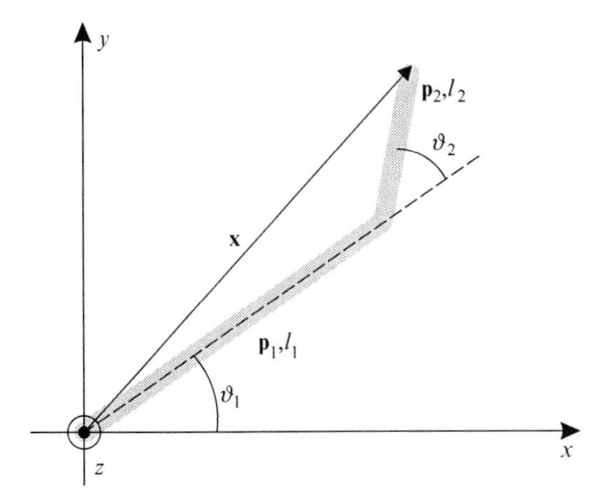

Fig. 4.1 Planar two-segment robot manipulator

trajectories. We shall first find the solution of direct kinematics for the trajectories. By differentiation we shall obtain the equations describing direct kinematics for the velocities. By simple matrix inversion inverse kinematics for the velocities can be computed. Let us now consider the planar two-segment robot manipulator shown in Figure 4.1.

The axis of rotation of the first joint is presented by the vertical z axis pointing out of the page. Vector $\mathbf{p}_1$ is directed along the first segment of the simple mechanism

$$\mathbf{p}_1 = l_1 \begin{bmatrix} \cos\vartheta_1 \\ \sin\vartheta_1 \end{bmatrix}. \tag{4.1}$$

Vector $\mathbf{p}_2$ is along with the second segment. Its components can be read from Figure 4.1

$$\mathbf{p}_2 = l_2 \begin{bmatrix} \cos(\vartheta_1+\vartheta_2) \\ \sin(\vartheta_1+\vartheta_2) \end{bmatrix}. \tag{4.2}$$

Vector $\mathbf{x}$ connects the origin of the coordinate frame with the robot end-point

$$\mathbf{x} = \mathbf{p}_1 + \mathbf{p}_2. \tag{4.3}$$

Vector $\mathbf{x}$ describes the position of the robot end-point

$$\mathbf{x} = \begin{bmatrix} x \\ y \end{bmatrix} = \begin{bmatrix} l_1\cos\vartheta_1 + l_2\cos(\vartheta_1+\vartheta_2) \\ l_1\sin\vartheta_1 + l_2\sin(\vartheta_1+\vartheta_2) \end{bmatrix}. \tag{4.4}$$

By defining the vector of joint angles

$$\mathbf{q} = \begin{bmatrix} \vartheta_1 & \vartheta_2 \end{bmatrix}^T, \tag{4.5}$$

the equation (4.4) can be written in the following shorter form

$$\mathbf{x} = \mathbf{k}(\mathbf{q}), \tag{4.6}$$

where $\mathbf{k}(\cdot)$ represents the equations of direct kinematics.

The relation between the velocities of the robot end-point and joint velocities is obtained by differentiation. The coordinates of the end-point are functions of the joint angles which in turn are functions of time

$$\begin{aligned} x &= x(\vartheta_1(t), \vartheta_2(t)) \\ y &= y(\vartheta_1(t), \vartheta_2(t)). \end{aligned} \tag{4.7}$$

By calculating the time derivatives of equations (4.7) and arranging them into matrix form, we can write

$$\begin{bmatrix} \dot{x} \\ \dot{y} \end{bmatrix} = \begin{bmatrix} \frac{\partial x}{\partial \vartheta_1} & \frac{\partial x}{\partial \vartheta_2} \\ \frac{\partial y}{\partial \vartheta_1} & \frac{\partial y}{\partial \vartheta_2} \end{bmatrix} \begin{bmatrix} \dot{\vartheta}_1 \\ \dot{\vartheta}_2 \end{bmatrix}. \tag{4.8}$$

For our two-segment robot manipulator we obtain the following expression

$$\begin{bmatrix} \dot{x} \\ \dot{y} \end{bmatrix} = \begin{bmatrix} -l_1 s1 - l_2 s12 & -l_2 s12 \\ l_1 c1 + l_2 c12 & l_2 c12 \end{bmatrix} \begin{bmatrix} \dot{\vartheta}_1 \\ \dot{\vartheta}_2 \end{bmatrix}. \tag{4.9}$$

The matrix, which is in our case of the second order, is called the Jacobian matrix $\mathbf{J}(\mathbf{q})$. The relation (4.9) can be written in short form as

$$\dot{\mathbf{x}} = \mathbf{J}(\mathbf{q})\dot{\mathbf{q}}. \tag{4.10}$$

In this way the problems of direct kinematics for the trajectories and velocities are solved. When solving the inverse kinematics, we calculate the joint angles from the known position of the robot end-point. Figure 4.2 shows only those parameters of the two-segment robot mechanism which are relevant for the calculation of the ϑ_2 angle. The cosine rule is used

$$x^2 + y^2 = l_1^2 + l_2^2 - 2 l_1 l_2 \cos(180^\circ - \vartheta_2).$$

The angle in the second joint of the two-segment manipulator is calculated as the inverse trigonometric function

$$\vartheta_2 = \arccos \frac{x^2 + y^2 - l_1^2 - l_2^2}{2 l_1 l_2}. \tag{4.11}$$

The angle in the first joint is calculated by the use of Figure 4.3. It is obtained as the difference of the angles α_1 and α_2

$$\vartheta_1 = \alpha_1 - \alpha_2.$$

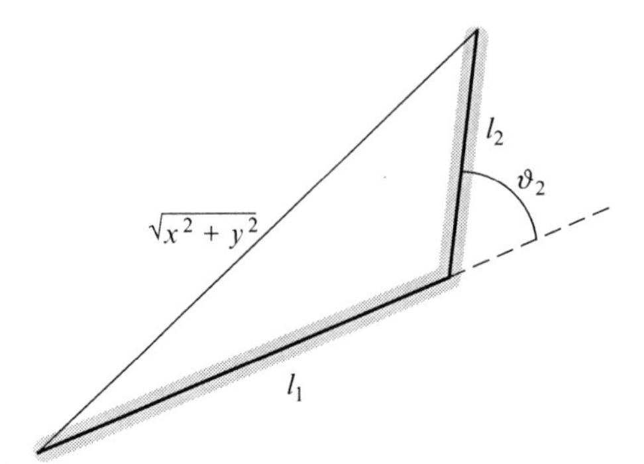

Fig. 4.2 Calculation of the ϑ_2 angle

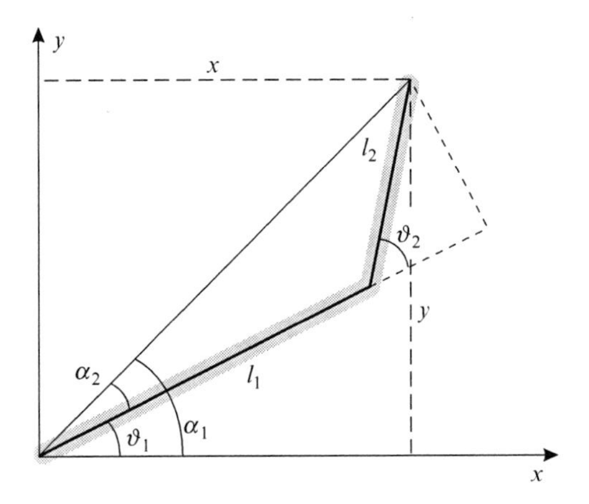

Fig. 4.3 Calculation of the ϑ_1 angle

The angle α_1 is obtained from the right-angle triangle made of horizontal x and vertical y coordinates of the robot end-point. The angle α_2 is obtained by elongating the triangle of Figure 4.2 into the right-angle triangle, as shown in Figure 4.3. Again we make use of the inverse trigonometric functions

$$\vartheta_1 = \arctan\left(\frac{y}{x}\right) - \arctan\left(\frac{l_2 \sin\vartheta_2}{l_1 + l_2\cos\vartheta_2}\right). \tag{4.12}$$

When calculating the ϑ_2 angle, we have two solutions, "elbow-up" and "elbow-down", as shown in Figure 4.4. A degenerate solution is represented by the end-point position $x = y = 0$ when both segments are of equal length $l_1 = l_2$. In this case $\arctan\left(\frac{y}{x}\right)$ is not defined. When the angle $\vartheta_2 = 180°$, the base of the simple two-segment mechanism can be reached at an arbitrary angle ϑ_1. When a point (x,y) lays out of the manipulator workspace, the problem of inverse kinematics cannot be solved.

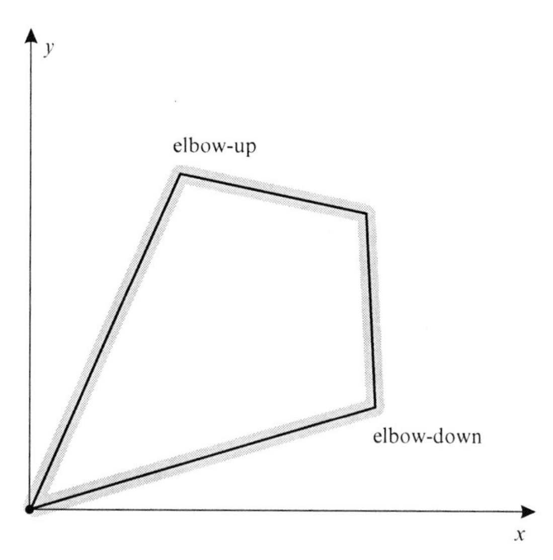

Fig. 4.4 Two solutions of inverse kinematics

The relation between the joint velocities and the velocity of the end-point is obtained by inverting the Jacobian matrix $\mathbf{J}(\mathbf{q})$

$$\dot{\mathbf{q}} = \mathbf{J}^{-1}(\mathbf{q})\dot{\mathbf{x}}. \tag{4.13}$$

The matrices of order 2×2 can be inverted as follows

$$\mathbf{A} = \begin{bmatrix} a & b \\ c & d \end{bmatrix} \qquad \mathbf{A}^{-1} = \frac{1}{ad - cb} \begin{bmatrix} d & -b \\ -c & a \end{bmatrix}.$$

For our two-segment manipulator we can write

$$\begin{bmatrix} \dot{\vartheta}_1 \\ \dot{\vartheta}_2 \end{bmatrix} = \frac{1}{l_1 l_2 s2} \begin{bmatrix} l_2 c12 & l_2 s12 \\ -l_1 c1 - l_2 c12 & -l_1 s1 - l_2 s12 \end{bmatrix} \begin{bmatrix} \dot{x} \\ \dot{y} \end{bmatrix}. \tag{4.14}$$

In general examples of robot manipulators, it is not necessary that the Jacobian matrix has the quadratic form. In this case, the so called pseudoinverse matrix $(\mathbf{J}\mathbf{J}^T)^{-1}$ is calculated. For a robot with six degrees of freedom the Jacobian matrix is quadratic, however after inverting, it becomes rather impractical. When the manipulator is close to singular poses (e.g. when angle ϑ_2 is close to zero for the simple two-segment robot), the inverse Jacobian matrix is ill defined. We shall make use of the Jacobian matrix when studying robot control.

At the end of the chapter let us make a short leap from robot kinematics to robot statics. Let us suppose that the end-point of the two-segment robot manipulator bumped into an obstacle (Figure 4.5). In this way the robot is producing a force against the obstacle. The horizontal component of the force acts in the positive direction of the x axis, while the vertical component is directed along the y axis. The force

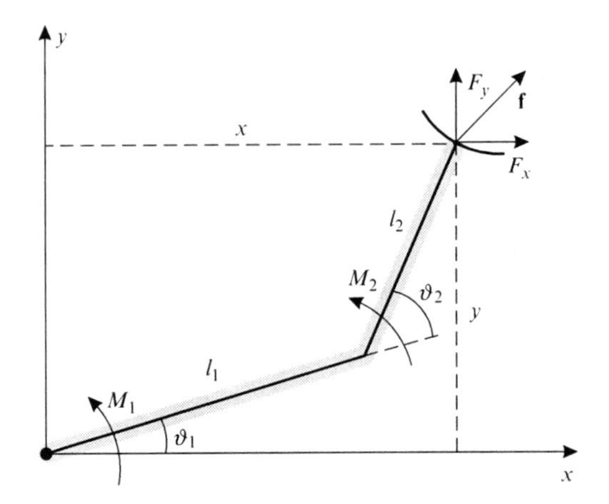

Fig. 4.5 Two-segment robot manipulator in contact with the environment

against the obstacle is produced by the motors in the robot joints. The motor of the first joint is producing the joint torque M_1, while M_2 is the torque in the second joint.

The positive directions of both joint torques are counter-clockwise. As the robot is not moving, the sum of the external torques equals zero. This means that the torque M_1 in the first joint is equal to the torque of the external force or it is equal to the torque that the manipulator exerts on the obstacle

$$M_1 = -F_x y + F_y x. \tag{4.15}$$

The end-point coordinates x and y, calculated by equations (4.4), are inserted into equation (4.15)

$$M_1 = -F_x(l_1 \sin\vartheta_1 + l_2 \sin(\vartheta_1 + \vartheta_2)) + F_y(l_1 \cos\vartheta_1 + l_2 \cos(\vartheta_1 + \vartheta_2)). \tag{4.16}$$

In a similar way the torque in the second joint is determined

$$M_2 = -F_x l_2 \sin(\vartheta_1 + \vartheta_2) + F_y l_2 \cos(\vartheta_1 + \vartheta_2). \tag{4.17}$$

Equations (4.16) and (4.17) can be written in matrix form

$$\begin{bmatrix} M_1 \\ M_2 \end{bmatrix} = \begin{bmatrix} -l_1 s1 - l_2 s12 & l_1 c1 + l_2 c12 \\ -l_2 s12 & l_2 c12 \end{bmatrix} \begin{bmatrix} F_x \\ F_y \end{bmatrix}. \tag{4.18}$$

The matrix in equation (4.18) is a transposed Jacobian matrix. The transposed matrix of order 2×2 has the following form

$$\mathbf{A} = \begin{bmatrix} a & b \\ c & d \end{bmatrix} \qquad \mathbf{A}^T = \begin{bmatrix} a & c \\ b & d \end{bmatrix}.$$

In this way we obtained an important relation between the joint torques and the forces at the robot end-effector

$$\tau = \mathbf{J}^T(\mathbf{q})\mathbf{f}, \tag{4.19}$$

where

$$\tau = \begin{bmatrix} M_1 \\ M_2 \end{bmatrix} \qquad \mathbf{f} = \begin{bmatrix} F_x \\ F_y \end{bmatrix}.$$

Equation (4.19) describes the robot statics. It will be used in the control of a robot which is in contact with the environment.

4.2 Workspace

The robot workspace consists of all points that can be reached by the robot end-point. It plays an important role when selecting an industrial robot for an anticipated task. It is our aim to describe an approach to determine the workspace of a chosen robot. We shall again consider the example of the simple planar two-segment robot with rotational joints. Our study of the robot workspace will in this way take place in a plane and we shall in fact deal with a working surface. Regardless of the constraints imposed by the plane we shall become aware of the most important characteristic properties of the robot workspaces. Industrial robots usually have the ability to turn around the first vertical joint axis. We shall therefore rotate the working surface around the vertical axis of the reference coordinate frame and thus obtain an idea of the realistic robot workspaces.

Let us consider the planar two-segment robot manipulator as shown in Figure 4.6. The rotational degrees of freedom are denoted as ϑ_1 and ϑ_2 and the lengths of the segments l_1 and l_2 will be considered equal. The coordinates of the robot end-point can be expressed with the following two equations

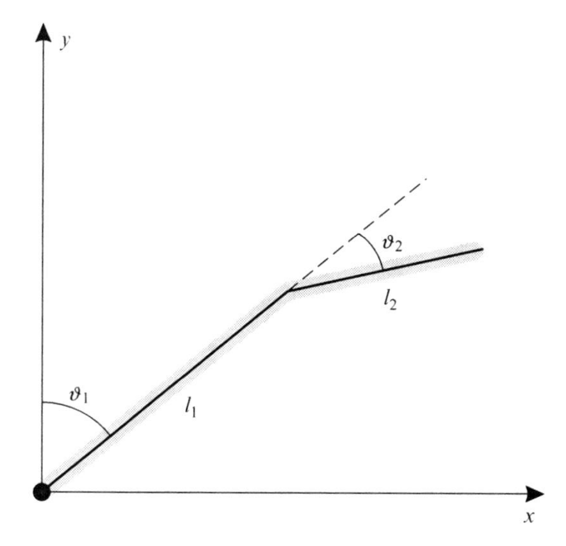

Fig. 4.6 Two-segment robot manipulator

$$x = l_1 \sin \vartheta_1 + l_2 \sin(\vartheta_1 + \vartheta_2) \tag{4.20}$$
$$y = l_1 \cos \vartheta_1 + l_2 \cos(\vartheta_1 + \vartheta_2).$$

If equations (4.20) are first squared and then summed, the equations of a circle are obtained

$$(x - l_1 \sin \vartheta_1)^2 + (y - l_1 \cos \vartheta_1)^2 = l_2^2 \tag{4.21}$$
$$x^2 + y^2 = l_1^2 + l_2^2 + 2 l_1 l_2 \cos \vartheta_2.$$

The first equation depends only on the angle ϑ_1, while only ϑ_2 appears in the second equation. The mesh of the circles plotted for different values ϑ_1 and ϑ_2 is shown in Figure 4.7. The first equation describes the circles which are in Figure 4.7 denoted

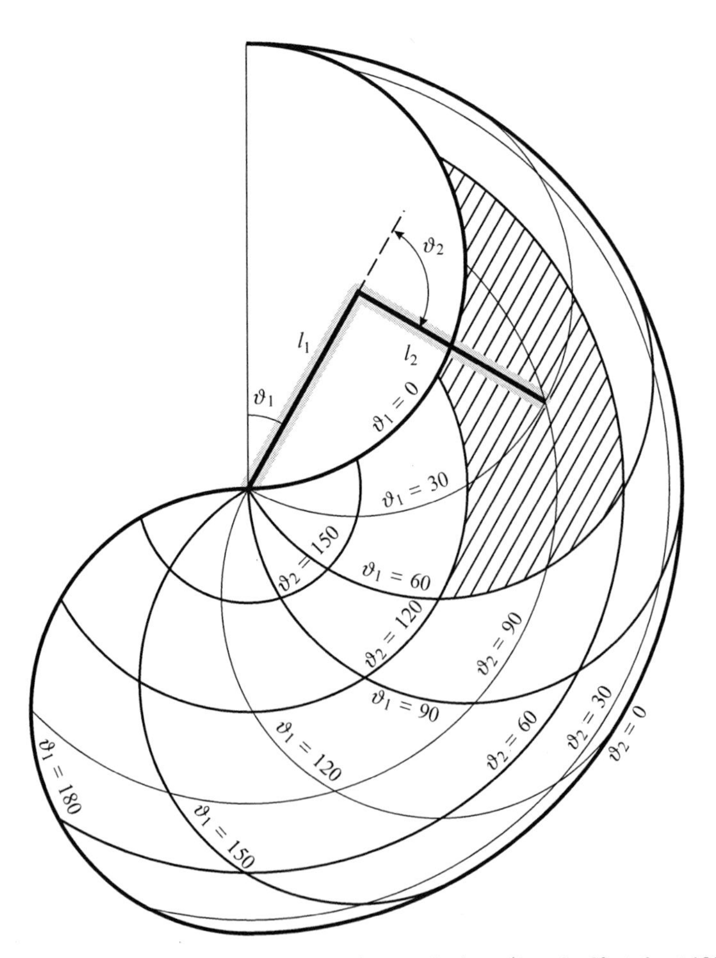

Fig. 4.7 Workspace of a planar two-segment robot manipulator ($l_1 = l_2$, $0° \leq \vartheta_1 \leq 180°$, $0° \leq \vartheta_2 \leq 180°$)

as $\vartheta_1 = 0^\circ, 30^\circ, 60^\circ, 90^\circ, 120^\circ, 150^\circ$, and 180°. Their radii are equal to the length of the second segment l_2, the centers of the circles depend on the angle ϑ_1 and travel along a circle with the center in the origin of the coordinate frame and with the radius l_1. The circles from the second equation have all their centers in the origin of the coordinate frame, while their radii depend on the lengths of both segments and the angle ϑ_2 between them.

The mesh in Figure 4.7 serves for simple graphical presentation of the working surface of a two-segment robot. It is not difficult to determine the working surface for the case when ϑ_1 and ϑ_2 vary in the full range from 0° to 360°. For the two-segment manipulator with equal lengths of both segments this is simply a circle with the radius $l_1 + l_2$. Much more irregular shapes of workspaces are obtained when the range of motion of the robot joints is constrained, as it is usually the case. Part of the working surface where ϑ_1 changes from 0° to 60° and ϑ_2 from 60° and 120° is in Figure 4.7 displayed as hatched.

When plotting the working surfaces of the two-segment manipulator we assumed that the lengths of both segments are equal. This assumption will be now supported by an adequate proof. It is not difficult to realize that the segments of industrial SCARA and anthropomorphic robots are of equal length. Let us consider a two-segment robot, where the second segment is shorter than the first one, while the angles ϑ_1 and ϑ_2 vary from 0° to 360° (Figure 4.8). The working area of such a manipulator is a ring with inner radius $R_i = l_1 - l_2$ and outer radius $R_o = l_1 + l_2$. It is our aim to find the ratio of the segments lengths l_1 and l_2 resulting in the largest working area at constant collective length of both segments R_o. The working area of the described two-segment robot manipulator is

$$A = \pi R_o^2 - \pi R_i^2. \tag{4.22}$$

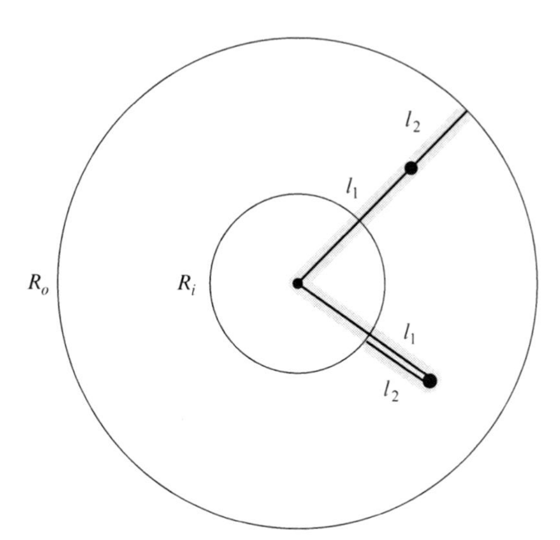

Fig. 4.8 Working area of two-segment manipulator with the second segment shorter

By inserting the expression for the inner radius in equation (4.22)

$$R_i^2 = (l_1 - l_2)^2 = (2l_1 - R_o)^2 \tag{4.23}$$

we can write

$$A = \pi R_o^2 - \pi(2l_1 - R_o)^2. \tag{4.24}$$

The derivative of the working area with respect to the segment length l_1 is equal to zero

$$\frac{\partial A}{\partial l_1} = 2\pi(2l_1 - R_o) = 0. \tag{4.25}$$

The solution is

$$l_1 = \frac{R_o}{2}, \tag{4.26}$$

giving

$$l_1 = l_2. \tag{4.27}$$

The largest working area of the two-segment mechanism occurs for equal lengths of both segments.

The area of the working surface depends on the segment lengths l_1 and l_2 and on the minimal and maximal values of the angles ϑ_1 and ϑ_2. When changing the ratios l_1/l_2 we can obtain various shapes of the robot working surface. The area of such working surface is always equal to the one shown in Figure 4.9. In the Figure 4.9 ϑ_1 means the difference between the maximal and minimal joint angle value $\vartheta_1 = (\vartheta_{1_{max}} - \vartheta_{1_{min}})$. The area of the working surface is the area of a ring segment

$$A = \frac{\vartheta_1 \pi}{360}(r_1^2 - r_2^2) \text{ for } \vartheta_1\,[^\circ]. \tag{4.28}$$

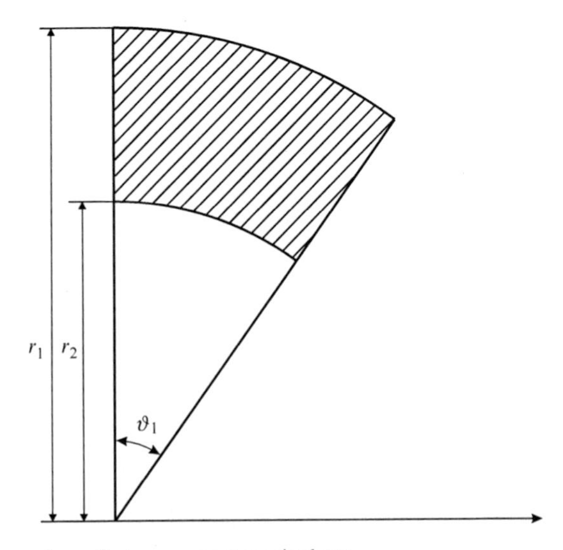

Fig. 4.9 Working surface of a two-segment manipulator

In equation (4.28) the radii r_1 and r_2 are obtained by the cosine rule

$$r_1 = \sqrt{l_1^2 + l_2^2 + 2l_1l_2\cos\vartheta_{2_{min}}} \qquad r_2 = \sqrt{l_1^2 + l_2^2 + 2l_1l_2\cos\vartheta_{2_{max}}}. \tag{4.29}$$

The area of the working surface is, in the same way as its shape, dependent on the ratio l_2/l_1 and the constraints in the joint angles. The angle ϑ_1 determines the position of the working surface with respect to the reference frame and has no influence on its shape. Let us examine the influence of the second angle ϑ_2 on the area of the working surface. We shall assume $l_1 = l_2 = 1$ and ϑ_1 changing from 30° to 60°. For equal ranges of the angle ϑ_2 (30°) and different values of $\vartheta_{2_{max}}$ and $\vartheta_{2_{min}}$ we obtain different values of the working areas

$$\begin{aligned}
0 \leq \vartheta_2 \leq 30 &\quad A = 0,07\\
30 \leq \vartheta_2 \leq 60 &\quad A = 0,19\\
60 \leq \vartheta_2 \leq 90 &\quad A = 0,26\\
90 \leq \vartheta_2 \leq 120 &\quad A = 0,26\\
120 \leq \vartheta_2 \leq 150 &\quad A = 0,19\\
150 \leq \vartheta_2 \leq 180 &\quad A = 0,07.
\end{aligned}$$

Until now, under the term workspace we were considering the so called reachable robot workspace. This means all the points in the robot surrounding that can be reached by the robot end-point. Many times the so called dexterous workspace is of greater importance. The dexterous workspace comprises all the points that can be reached at an arbitrary orientation of the robot end-effector. This workspace is always smaller than the reachable workspace. The dexterous workspace is larger when the last segment (end-effector) is shorter. The reachable and the dexterous workspaces of a two-segment robot with the end-effector are shown in Figure 4.10.

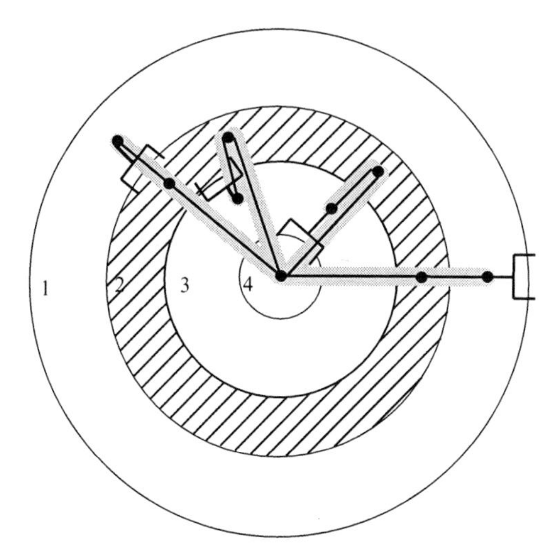

Fig. 4.10 Reachable and dexterous workspace of a two-segment manipulator with end-effector

The second and the third circle are obtained when the robot end-effector is oriented towards the area constrained by the two circles. These two circles represent the limits of the dexterous workspace. The first and the fourth circle constrain the reachable workspace. The points between the first and the second and the third and the fourth circle cannot be reached with an arbitrary orientation of the end-effector.

With robots having more than three joints, the described graphical approach is not appropriate. We make use of numerical methods and computer algorithms.

4.3 Dynamics

In contrast to kinematics, dynamics represents the part of mechanics, which is interested into the forces and torques which are producing the motion of a mechanism. Dynamic analysis has a double role in robotics. We apply it both in modeling and control of robot mechanisms. Before a robot is introduced into an industrial process, the robot task is simulated in a virtual environment by using special software for computer design of the robot cells. Such software is usually developed by robot producers. Its central part is the dynamic model of the robot which we will get to know in a simplified version in this chapter. In the virtual robot environment, also the material handling devices, such as containers, pallets and conveyors, interacting with the robot, are included. Dynamic equations of the robot movements also offer relevant information for the design of robot controllers. By the use of a dynamic model we can compute in real time the motor torques required to produce the desired motion of the robot. The dynamic robot analysis enables us to consider properly

- The torques necessary to compensate the gravity forces of robot segments
- The differences in moments of inertia occurring during the robot motion
- Dynamic couplings caused by simultaneous movements of all robot segments

In a similar way as in kinematics, we distinguish between the direct and inverse dynamics. When solving the inverse dynamics problem, the joint motion is known, while we are calculating the torques and forces producing this motion. In direct dynamics we calculate the joint trajectories, velocities and accelerations from the known forces and torques produced by the actuators in the robot joints. We will be mainly interested in the inverse dynamic problem, i.e. calculation of the forces and torques in the robot joints. Later we can, by the use of the model of the actuator (electrical or hydraulic) and the reducer, calculate the voltage, which must be applied to each particular motor in order to obtain the desired motion of the robot.

The rather complex dynamic analysis of a robot mechanism will be explained with the help of the two-segment robot mechanism shown in Figure 4.11. The motion of the robot manipulator with two rotational joints occurs in the vertical plane. Both segments are of equal length l. The dynamic model will be simplified in such

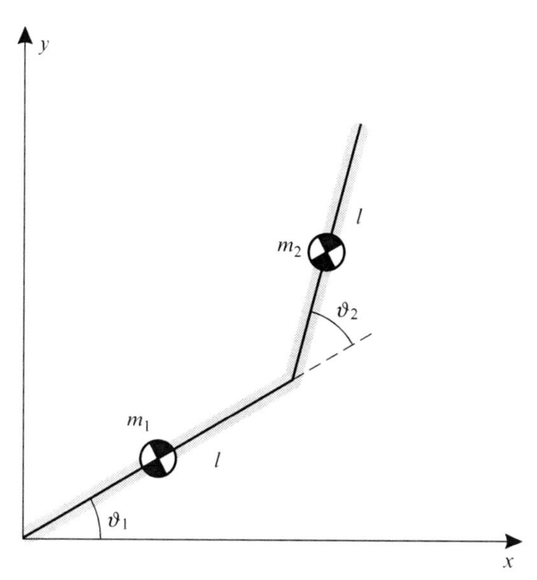

Fig. 4.11 Planar two-segment robot manipulator with two rotational joints

a way that we shall assume that the whole mass of each segment is concentrated in its center of mass. The mass of the first segment is m_1 and m_2 is the mass of the second segment. The dynamic analysis of such a pair of segments is interesting because it appears both in the anthropomorphic and in the SCARA robot structures. The joint trajectories are denoted by the angles ϑ_1 and ϑ_2. The simple two-segment robot manipulator is placed into the fixed reference frame whose z axis is aligned with the axis of the first joint.

Let us first calculate the torque M_2, which is produced by the actuator in the second joint of the simple robot mechanism, in order to achieve the desired trajectory, velocity and acceleration of the robot end-point. From Figure 4.12 we see that the position, velocity and acceleration of the center of mass of the second segment are given by

$$\begin{aligned} \mathbf{r} &= \mathbf{r}_1 + \mathbf{r}_2 \\ \mathbf{v} = \dot{\mathbf{r}} &= \dot{\mathbf{r}}_1 + \dot{\mathbf{r}}_2 \\ \mathbf{a} = \ddot{\mathbf{r}} &= \ddot{\mathbf{r}}_1 + \ddot{\mathbf{r}}_2. \end{aligned} \tag{4.30}$$

The motion of the mass m_2 is given by Newton's law

$$\sum_i \mathbf{F}_i = m_2 \mathbf{a}. \tag{4.31}$$

In addition to the force of gravity, the mass m_2 is acted upon by the force $\mathbf{F_2}$, transmitted by the massless segment $\mathbf{r}_2$. Therefore

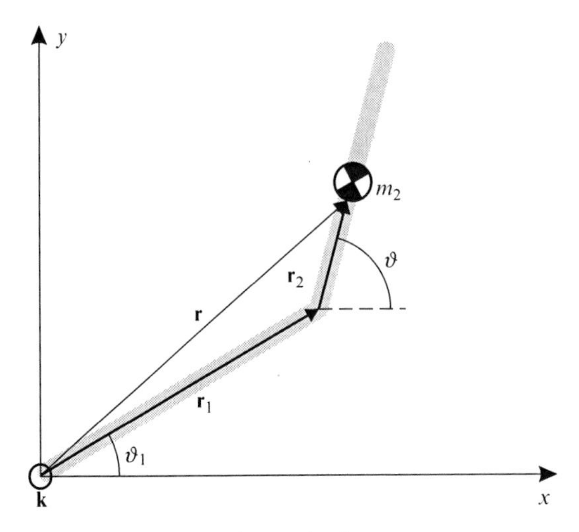

Fig. 4.12 Model of planar two-segment robot mechanism

$$\mathbf{F}_2 + m_2\mathbf{g} = m_2(\ddot{\mathbf{r}}_1 + \ddot{\mathbf{r}}_2). \tag{4.32}$$

As the robot segments $\mathbf{r}_1$ and $\mathbf{r}_2$ are rigid, the second derivatives can be written as

$$\begin{aligned} \ddot{\mathbf{r}}_1 &= -\dot{\vartheta}_1^2\mathbf{r}_1 + \ddot{\vartheta}_1(\mathbf{k}\times\mathbf{r}_1) \\ \ddot{\mathbf{r}}_2 &= -\dot{\vartheta}^2\mathbf{r}_2 + \ddot{\vartheta}(\mathbf{k}\times\mathbf{r}_2). \end{aligned} \tag{4.33}$$

Here, unit vector $\mathbf{k}$ is perpendicular to the plane x, y, while the angle ϑ represents the sum of the angles ϑ_1 and ϑ_2, so

$$\begin{aligned} \dot{\vartheta} &= \dot{\vartheta}_1 + \dot{\vartheta}_2 \\ \ddot{\vartheta} &= \ddot{\vartheta}_1 + \ddot{\vartheta}_2. \end{aligned} \tag{4.34}$$

The torque in the second joint is thus obtained from equation (4.32) as

$$\mathbf{M}_2 = \mathbf{r}_2\times\mathbf{F}_2 = \mathbf{r}_2\times m_2\ddot{\mathbf{r}}_1 + \mathbf{r}_2\times m_2\ddot{\mathbf{r}}_2 - \mathbf{r}_2\times m_2\mathbf{g}. \tag{4.35}$$

By inserting expressions for (4.33) and (4.34) into equation (4.35) we obtain

$$\begin{aligned} \frac{\mathbf{M}_2}{m_2} =& -\dot{\vartheta}_1^2(\mathbf{r}_2\times\mathbf{r}_1) + \ddot{\vartheta}_1\left[\mathbf{r}_2\times(\mathbf{k}\times\mathbf{r}_1)\right] - \dot{\vartheta}^2(\mathbf{r}_2\times\mathbf{r}_2) \\ &+ \ddot{\vartheta}\left[\mathbf{r}_2\times(\mathbf{k}\times\mathbf{r}_2)\right] - \mathbf{r}_2\times\mathbf{g}. \end{aligned} \tag{4.36}$$

Considering the properties of the vector product of two and three vectors we can write

$$\frac{\mathbf{M}_2}{m_2} = \dot{\vartheta}_1^2 r_1 r_2 \sin\vartheta_2\mathbf{k} + \ddot{\vartheta}_1 r_1 r_2\cos\vartheta_2\mathbf{k} + (\ddot{\vartheta}_1+\ddot{\vartheta}_2)r_2^2\mathbf{k} + r_2 g\cos(\vartheta_1+\vartheta_2)\mathbf{k}. \tag{4.37}$$

The torque produced by the actuator in the second joint is obtained by inserting $r_1 = l$ and $r_2 = \frac{l}{2}$ into equation (4.37)

$$M_2 = \left[\frac{1}{4}m_2 l^2 + \frac{1}{2}m_2 l^2 c2\right]\ddot{\vartheta}_1 + \frac{1}{4}m_2 l^2 \ddot{\vartheta}_2 + \frac{1}{2}m_2 l^2 s2 \dot{\vartheta}_1^2 + \frac{1}{2}m_2 g l c12. \quad (4.38)$$

Let us examine the physical meaning of the terms in equation (4.38). The terms including the angular accelerations $\ddot{\vartheta}_1$ and $\ddot{\vartheta}_2$ are called inertial terms. In the first term there is mass m_2 of the second segment and acceleration $\ddot{\vartheta}_1$ of the first joint. This term describes the coupling of inertial effects in the two-segment manipulator. In the second term both mass and acceleration belong to the second segment and joint respectively. The third term is, due to the acceleration $l\dot{\vartheta}_1^2$, called centrifugal, while the fourth term is gravitational.

More computation is necessary for determining the torque, produced by the actuator in the first joint. We shall make use of the relation between the total torque of external forces and the time derivative of the angular momentum

$$\sum_i \mathbf{M}_i = \frac{\mathrm{d}}{\mathrm{d}t}\sum_i \Gamma_i. \quad (4.39)$$

We first write the sum of the torques produced by the external forces

$$\sum_i \mathbf{M}_i = \mathbf{M}_1 + \left(\frac{\mathbf{r}_1}{2}\right) \times m_1\mathbf{g} + (\mathbf{r}_1 + \mathbf{r}_2) \times m_2\mathbf{g}. \quad (4.40)$$

In the equation (4.40) $\mathbf{M}_1$ denotes the torque produced by the motor in the first joint, while the other two terms are due to gravity. The angular momentum of the mass m_1 is

$$\Gamma_1 = \left(\frac{\mathbf{r}_1}{2}\right) \times m_1\mathbf{v}_1, \quad (4.41)$$

where the velocity $\mathbf{v}_1$ is

$$\mathbf{v}_1 = \dot{\vartheta}_1\left(\mathbf{k} \times \left(\frac{\mathbf{r}_1}{2}\right)\right). \quad (4.42)$$

The angular momentum of mass m_2 is

$$\Gamma_2 = (\mathbf{r}_1 + \mathbf{r}_2) \times m_2\mathbf{v}_2, \quad (4.43)$$

where the velocity $\mathbf{v}_2$ is

$$\mathbf{v}_2 = \dot{\mathbf{r}} = \dot{\mathbf{r}}_1 + \dot{\mathbf{r}}_2 = \dot{\vartheta}_1(\mathbf{k} \times \mathbf{r}_1) + (\dot{\vartheta}_1 + \dot{\vartheta}_2)(\mathbf{k} \times \mathbf{r}_2). \quad (4.44)$$

After rearranging equations (4.39)–(4.44) and taking into account $r_1 = l$ and $r_2 = \frac{l}{2}$ the torque in the second joint is obtained in the following form

$$
\begin{aligned}
M_1 = & \left[\frac{1}{4}m_1l^2 + \frac{5}{4}m_2l^2 + m_2l^2c2\right]\ddot{\vartheta}_1 + \\
& + \left[\frac{1}{4}m_2l^2 + \frac{1}{2}m_2l^2c2\right]\ddot{\vartheta}_2 - \\
& - m_2l^2s2\dot{\vartheta}_1\dot{\vartheta}_2 - \frac{1}{2}m_2l^2s2\dot{\vartheta}_2^2 + \\
& + \frac{1}{2}m_1glc1 + m_2glc1 + \frac{1}{2}m_2glc12.
\end{aligned}
\tag{4.45}
$$

As in equation (4.38), here again we encounter the inertial, centrifugal and three gravitational terms. There is also an additional term characterized by the product of angular velocities. It is called the Coriolis term. This torque results from the variation of the angular momentum of the mass m_2 which occurs because of the variation of the moment of inertia around the axis of rotation.

The torques in the robot joints can be written in the following general form

$$
\tau = \mathbf{B}(\mathbf{q})\ddot{\mathbf{q}} + \mathbf{C}(\mathbf{q},\dot{\mathbf{q}})\dot{\mathbf{q}} + \mathbf{g}(\mathbf{q}). \tag{4.46}
$$

In equation (4.46) the vector τ unites the torques of both actuators

$$
\tau = \begin{bmatrix} M_1 \\ M_2 \end{bmatrix}.
$$

Vectors $\mathbf{q}$, $\dot{\mathbf{q}}$, and $\ddot{\mathbf{q}}$ belong to the joint trajectories, velocities and accelerations respectively. For the two-segment robot we have

$$
\mathbf{q} = \begin{bmatrix} \vartheta_1 \\ \vartheta_2 \end{bmatrix} \qquad \dot{\mathbf{q}} = \begin{bmatrix} \dot{\vartheta}_1 \\ \dot{\vartheta}_2 \end{bmatrix} \qquad \ddot{\mathbf{q}} = \begin{bmatrix} \ddot{\vartheta}_1 \\ \ddot{\vartheta}_2 \end{bmatrix}.
$$

The first term of equation (4.46) is the inertial term. In our case we are dealing with the following inertial matrix $\mathbf{B}(\mathbf{q})$

$$
\mathbf{B}(\mathbf{q}) = \begin{bmatrix} \frac{1}{4}m_1l^2 + \frac{5}{4}m_2l^2 + m_2l^2c2 & \frac{1}{4}m_2l^2 + \frac{1}{2}m_2l^2c2 \\ \frac{1}{4}m_2l^2 + \frac{1}{2}m_2l^2c2 & \frac{1}{4}m_2l^2 \end{bmatrix}.
$$

The second term in the equation (4.46) is called the Coriolis term and includes velocity and centrifugal effects. For the two-segment robot we have the following matrix

$$
\mathbf{C}(\mathbf{q},\dot{\mathbf{q}}) = \begin{bmatrix} -m_2l^2s2\dot{\vartheta}_2 & -\frac{1}{2}m_2l^2s2\dot{\vartheta}_2 \\ \frac{1}{2}m_2l^2s2\dot{\vartheta}_1 & 0 \end{bmatrix}.
$$

The gravitational column $\mathbf{g}(\mathbf{q})$ has in our case the following form

$$
\mathbf{g}(\mathbf{q}) = \begin{bmatrix} \frac{1}{2}m_1glc1 + m_2glc1 + \frac{1}{2}m_2glc12 \\ \frac{1}{2}m_2glc12 \end{bmatrix}.
$$

Chapter 5
Robot sensors

The human sensory system encompasses sensors of vision and hearing, kinesthetic sensors (movement, force and touch), sensors of taste and smell. These sensors deliver input signals to the brain which, on the basis of sensory information, builds its own image of the environment and takes decisions for further actions. Similar requirements are valid also for robot mechanisms. However, because of the complexity of human sensing, the robot sensing is limited to fewer sensors.

The use of sensors is of crucial importance for efficient and accurate robot operation. In general the robot sensors can be divided into: (1) proprioceptive sensors assessing the internal states of the robot mechanism (positions, velocities and torques in the robot joints) and (2) exteroceptive sensors delivering to the controller the information about the robot environment (force, tactile, proximity and distance sensors, robot vision).

5.1 Principles of sensing

In general, sensors convert the measured physical variable into an electric signal which can be in a digital form assessed by the computer. In robotics we are predominantly interested in the following variables: position, velocity, force and torque. By the use of special transducers these variables can be converted into electric signals, such as voltage, current, resistance, capacity or inductivity. Based on the principle of conversion the sensors can be divided into:

- Electric sensors – the physical variable is directly transformed into an electrical signal; such sensors are for example potentiometers or strain gauges
- Electromagnetic sensors – use the magnetic field for the purposes of physical variable conversion; an example is tachometer
- Optical sensors – use light when converting the signals; an example of such a sensor is the optical encoder

T. Bajd et al., *Robotics*, Intelligent Systems, Control and Automation: Science and Engineering 43, DOI 10.1007/978-90-481-3776-3_5,

5.2 Sensors of movement

Typical sensors of robot movements are potentiometers, optical encoders and tachometers. They all measure the robot movements inside the robot joint. It is important, however, where in the joint to place the sensor of movement and how to measure the motion parameters.

5.2.1 *Placing of sensors*

Let us first consider a sensor of angular displacement. It is our aim to measure the angle in a robot joint which is actuated by a motor through a reducer with the reduction ratio k_{ri}. Using a reducer we decrease the joint angular velocity by the factor k_{ri} with respect to the angular velocity of the motor. In the same time the joint torque is increased by the same factor. It is important whether the sensor of movement is placed before or behind the reducer. The choice depends on the task requirements and the sensor used. In an ideal case we mount the sensor before the reducer (on the side of the motor) as shown in Figure 5.1. In this way we measure directly the rotations of the motor. The sensor output must be then divided by the reduction ratio, in order to obtain the joint angle.

Let us denote by ϑ_i the angular position of the ith joint, ϑ_{mi} as the angular position of the corresponding motor and k_{ri} the reduction ratio of the ith reducer. When the sensor is placed before the reducer, its output is equal to the angle ϑ_{mi}. The variable which we need for control purposes is the joint angle ϑ_i, which is determined by the ratio

$$\vartheta_i = \frac{\vartheta_{mi}}{k_{ri}}. \tag{5.1}$$

By differentiating the equation (5.1) with respect to ϑ_{mi} we have

$$\frac{d\vartheta_i}{d\vartheta_{mi}} = \frac{1}{k_{ri}} \quad \text{thus} \quad d\vartheta_i = \frac{1}{k_{ri}} d\vartheta_{mi}, \tag{5.2}$$

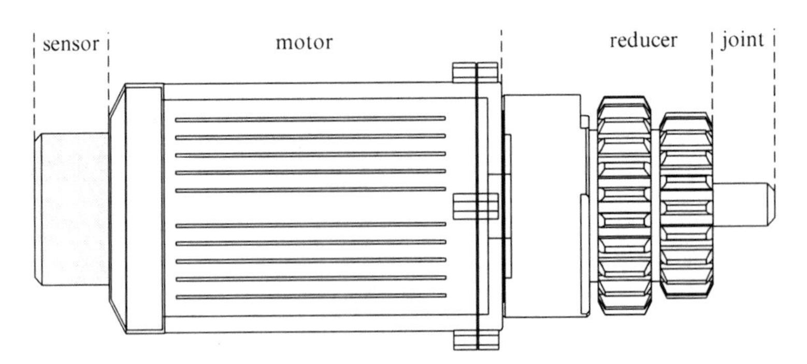

Fig. 5.1 Mounting of the sensor of movement before the reducer

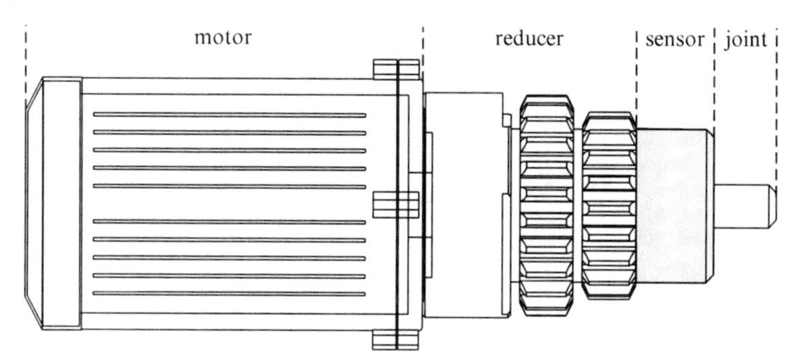

Fig. 5.2 Mounting of the sensor of movement behind the reducer

which means that the sensor measurement error is reduced by the factor k_{ri}. The advantage of the placement of the sensor before the reducer is in getting more accurate information about the joint angular position.

Another sensor mounting possibility is shown in Figure 5.2. Here, the sensor is mounted behind the reducer. In this way the movements of the joint are measured directly. The quality of the control signal is decreased, as the sensor measurement error, which is now not reduced, directly enters the joint control loop. As the range of motion of the joint is by the factor k_{ri} smaller than that of the motor, sensors with smaller range of motion can be used. In some cases we cannot avoid mounting of the motion sensor into the joint axis. It is important, therefore, that we are aware of the deficiency of such a placement.

5.2.2 Potentiometer

Figure 5.3 presents a model of a rotary potentiometer and its components. The potentiometer consists of two parts: (1) resistive winding and (2) movable wiper. The potentiometer represents a contact measuring method, because the wiper slides along the circular resistive winding.

Potentiometers are generally placed behind the reducer in such a way that the potentiometer axis is coupled to the joint axis. Let us suppose that point B represents the reference position of the potentiometer belonging to the ith joint. The resistance of the potentiometer along the winding $\widehat{AB}$ equals R, while r represents the resistance of the $\widehat{CB}$ part of the winding. The angle of the wiper with respect to the reference position B is denoted by ϑ_i (in radians). When the resistance along the circular winding of the potentiometer is uniform and the distance between the points A and B is negligible, we have the following equation

$$\frac{r}{R} = \frac{\widehat{CB}}{\widehat{AB}} = \frac{\vartheta_i}{2\pi}. \tag{5.3}$$

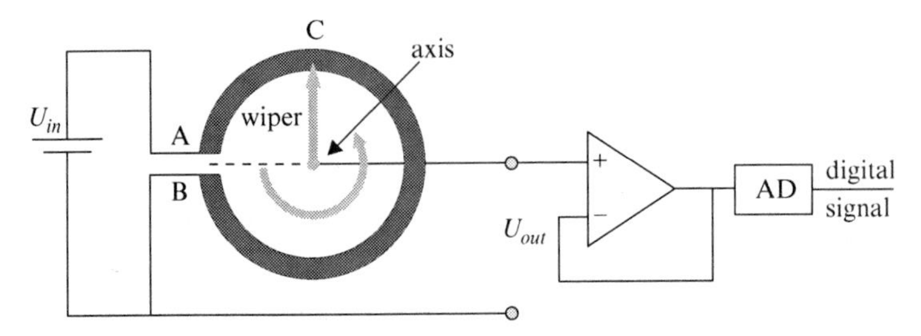

Fig. 5.3 The model of a potentiometer

Let us suppose that the potentiometer is supplied by the voltage U_{in}. The output voltage measured on the wiper is equal to

$$\frac{U_{out}}{U_{in}} = \frac{r}{R} = \frac{\vartheta_i}{2\pi}, \tag{5.4}$$

or

$$U_{out} = \frac{U_{in}}{2\pi}\vartheta_i. \tag{5.5}$$

By measuring the output voltage U_{out}, the angular position ϑ_i is determined.

5.2.3 *Optical encoder*

The contact measurement approach to the robot joint angle by using potentiometers has several deficiencies. The most important is the relatively short life time because of its wearing out. In addition, its most adequate placement is directly in the joint axis (behind the reducer) and not on the motor axis (before the reducer). The most widely used sensors of movements in robotics are therefore optical encoders providing contact-less measurement.

The optical encoder is based on the transformation of the joint movement into a series of light pulses, which are further converted into electric pulses. In order to generate the light pulses, a light source is needed, usually represented by a light emitting diode. The conversion of light into electric pulses is performed by the use of a phototransistor or a photodiode, converting light into electrical current. The model of an optical encoder assessing the joint angular position is presented in Figure 5.4. It consists of a light source with lens, light detector and a rotating disc with slots, which is connected to either motor or joint axis. On the rotating disc there is a track of slots and interspaces, which alternately either transfer or block the light from the light emitting diode to the phototransistor. The logical output of the sensor is high when the light goes through the slot and hits the phototransistor on the other side of the rotating plate. When the path between the light emitting diode and the phototransistor is blocked by the interspace between two slots, the logical output is low.

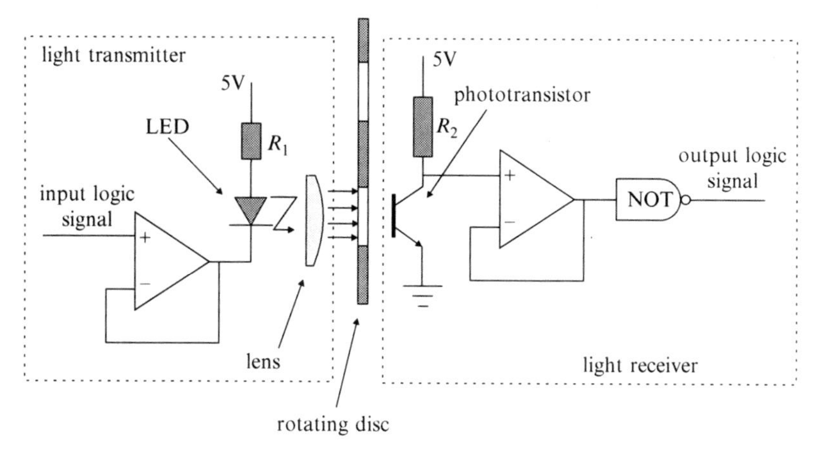

Fig. 5.4 The model of optical encoder

The optical encoders are divided into absolute and incremental. In the further text we shall learn about their most important characteristic properties.

5.2.3.1 Absolute encoders

The absolute optical encoder is a device which measures the absolute angular position of a joint. Its output is a digital signal. In a digital system each logical signal line represents one bit of information. When connecting all the bits of the absolute encoder into a single logical state variable, the number of all possible logical states determines the number of all absolute angular positions that can be measured by the absolute encoder.

Let us suppose that we wish to measure the angular rotation of 360° with the resolution of 0.1°. The absolute encoder must distinguish between 3,600 different logical states, which means that we need at least 12 bits to assess the joint angles with the required resolution. With 12 bits we can represent 4,096 logical states. An important design parameter of the absolute encoders is therefore the number of logical states, which depends on the task requirements and the placement of the encoder (before or behind the reducer). When the encoder is placed before a reducer with the reduction ratio k_{ri}, the resolution of the angle measurement will be increased by the factor k_{ri}. When the encoder is behind the reducer, the necessary resolution of the encoder is directly determined by the required resolution of the joint angle measurement. All logical states must be uniformly engraved into the rotating disc of the encoder. An example of absolute encoder with 16 logical states is shown in Figure 5.5. The 16 logical states can be represented by 4 bits. All 16 logical states are engraved into the surface of the rotating disc. The disc is in the radial direction divided into four tracks representing the 4 bits. Each track is divided into 16 segments corresponding to the logical states. As the information about the angular

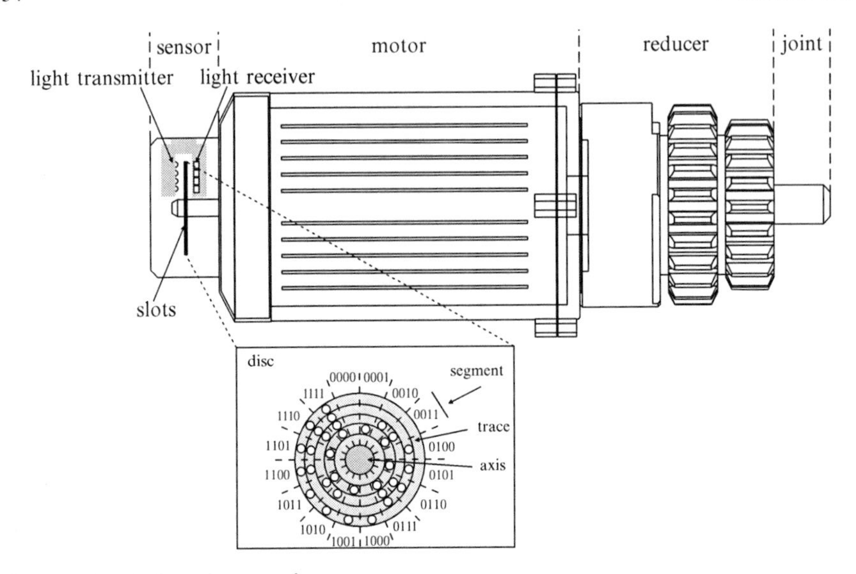

Fig. 5.5 Model of absolute encoder

displacement is represented by four bits, we need four pairs of light emitting diodes and phototransistors (one pair for each bit). With the rotation of the disc, which is connected to either motor or joint axis, the output signal will change between 0000 and 1111 for the positive direction of rotation and from 1111 to 0000 for the negative rotation. The absolute encoder does not determine only the angular position of the joint but also the direction of rotation.

5.2.3.2 Incremental encoders

In contrast to absolute encoders, the incremental encoders only supply the information about the changes in angular joint position. The advantages of the incremental encoders as compared to the absolute encoders are their simplicity, smaller dimensions and most importantly low cost. This can be achieved by lowering the number of the tracks on the rotating disc to only a single track. Instead of having as much tracks as the number of the bits necessary for the representation of all required logical states, we have now only one track with even graduation of the slots along the rim of the disc. Figure 5.6 shows a model of an incremental encoder. A single track only requires a single pair of light emitting diode and phototransistor (optical pair). During rotation of the encoded disc a series of electrical pulses is generated. The measurement of the joint displacement is based on counting of these pulses. Their number is proportional to the robot joint displacement. The incremental encoder shown in Figure 5.6 generates eight pulses during each rotation. The resolution of this encoder is

$$\Delta\vartheta = \frac{2\pi}{8} = \frac{\pi}{4}. \tag{5.6}$$

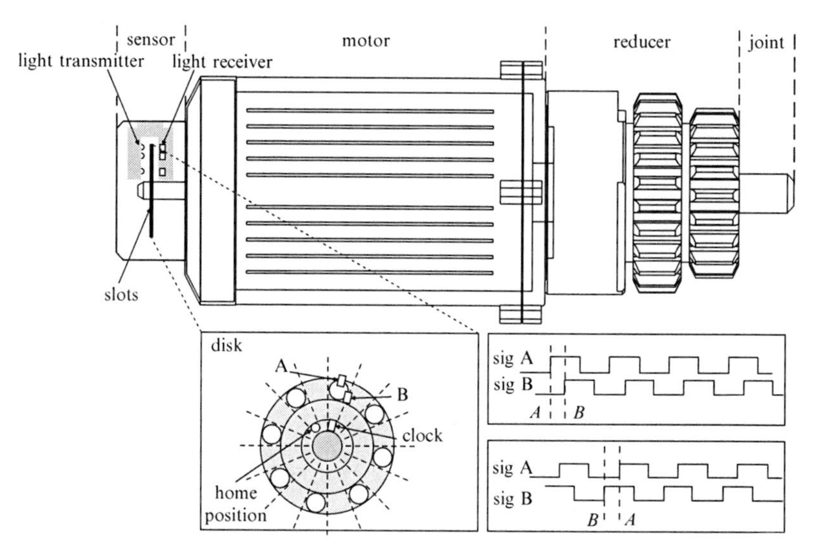

Fig. 5.6 Model of incremental encoder. The series of pulses for positive (above) and negative (below) direction of rotation

By increasing the number of the slots on the disc, the resolution of the encoder is increased. By denoting the number of the slots as n_c, a general equation for the encoder resolution can be written

$$\Delta\vartheta = \frac{2\pi}{n_c}. \tag{5.7}$$

The encoder with one single track is only capable of assessing the change in the joint angular position. It can not provide the information about the direction of rotation and the absolute joint position. If we wish to apply the incremental encoders in robot control, we must determine: (1) the home position representing the reference for the measurement of the change in the joint position and (2) the direction of rotation.

The problem of the home position is solved by adding an additional reference slot on the disc. This reference slot is displaced radially with respect to the slotted track measuring the angular position. For detection of the home position, an additional optical pair is needed. When searching for the reference slot, the robot is programmed to move with low velocity as long as the reference slot or the end position of the joint range of motion is reached. In the latter case the robot moves in the opposite direction towards the reference slot.

The problem of determining the direction of rotation is solved by another pair of light emitting diode and phototransistor. This additional optical pair is tangentially and radially displaced from the first optical pair as shown in Figure 5.6. When the disc is rotating, two signals are obtained, which are, because of the displacement of the optical pairs, shifted in phase. This shift in phase occurs because each slot on the disc first reaches the first optical pair and after a short delay also the second pair.

The optical components are usually placed in such a way that the phase shift of $\pi/2$ is obtained between the two signals. During the rotation in clockwise direction the signal B is phase-lagged for $\pi/2$ behind the signal A. During counter clockwise rotation the signal B is in phase-lead of $\pi/2$ with respect to the signal A (Figure 5.6). On the basis of the phase shifts between signals A and B, the direction of the encoder rotation can be determined.

5.2.4 *Tachometer*

The signal of the joint velocity can be obtained by numerical differentiation of the position signal. Nevertheless, direct measurement of the joint velocity with the help of a tachometer is often used in robotics. The reason is the noise introduced by numerical differentiation, which greatly affects the quality of the robot control.

Tachometers can be divided into: (1) direct current (DC) and (2) alternate current (AC) tachometers. In robotics mostly the simpler DC tachometers are used. The working principle is based on a DC generator whose magnetic field is provided by permanent magnets. As the magnetic field is constant, the tachometer output voltage is proportional to the angular velocity of the rotor. Because of the use of commutator in the DC tachometers, a slight ripple appears in the output voltage, which cannot be entirely filtered out. This deficiency, together with other imperfections, is avoided by the use of AC tachometers.

5.3 Force sensors

The sensors considered so far provide information about robot motions. They enable closing of the position and velocity control loop. In some robot tasks the contact of the end-effector with the environment is required. In these cases we must, besides the position, measure also the contact forces. In the simplest case the force measurement enables disconnection of the robot when the contact force exceeds a predetermined safety limit. In a more sophisticated case we use force sensors for control of the force between the robot end-effector and the environment. It is therefore not difficult to realize that the force sensor is placed into the robot wrist and is therefore often called the wrist sensor.

Strain gauges are usually used for the force measurements. The strain gauge is attached to an elastic beam which is deformed under the stress caused by the applied force. The strain gauge therefore behaves as a variable resistor whose resistance changes proportionally to its deformation. The wrist sensor must not influence the interaction of the robot with the environment. This means that the wrist sensor must be sufficiently rigid. The robot wrist sensors are usually designed as shown in Figure 5.7. The structure of the sensor is based on three components: (a) rigid inner ring which is in contact with the robot end-effector, (b) rigid outer ring which is in contact with the robot environment and (c) elastic beams interconnecting the outer

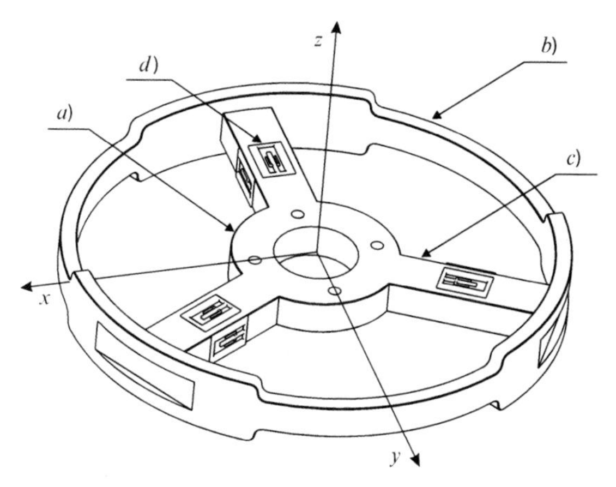

Fig. 5.7 Model of the force and torque sensor: (**a**) rigid ring which is in contact with the robot end-effector, (**b**) rigid ring which is in contact with the robot environment, (**c**) elastic beams and (**d**) strain gauge

and the inner ring. During contact of the robot with the environment, the beams are deformed by the external forces which causes a change in the resistance of the strain gauges.

The vector of the forces and torques acting at the robot end-effector is in the three-dimensional space represented by six elements, three forces and three torques. The rectangular cross-section of a beam, as shown in Figure 5.7, enables the measurement of deformations in two directions. In order to be able to measure the six elements of the force and torque vector, at least three beams, which are not collinear, are necessary. There are two strain gauges attached to the perpendicular surfaces of each beam. Having six strain gauges, there are six variable resistances R_1, R_2, R_3, R_4, R_5 and R_6. As the consequence of the external forces and torques, changes in the resistances ΔR_1, ΔR_2, ΔR_3, ΔR_4, ΔR_5 and ΔR_6 occur. The small changes in the resistance are, by the use of the Wheatstone bridge, converted into voltage signals (Figure 5.8). To each of the six variable resistors $\{R_1 \ldots R_6\}$ three additional resistors are added. The three resistors are, together with the strain gauge, connected into the measuring bridge. The bridge is supplied with the U_{in} voltage, while the output voltage U_{out} is determined by the difference $U_1 - U_2$. The U_1 voltage is

$$U_1 = \frac{R_{i,2}}{R_{i,1} + R_{i,2}} U_{in}, \tag{5.8}$$

while the U_2 voltage is

$$U_2 = \frac{R_i}{R_i + R_{i,3}} U_{in}. \tag{5.9}$$

The output voltage is equal to

$$U_{out} = \left(\frac{R_{i,2}}{R_{i,1} + R_{i,2}} - \frac{R_i}{R_i + R_{i,3}} \right) U_{in}. \tag{5.10}$$

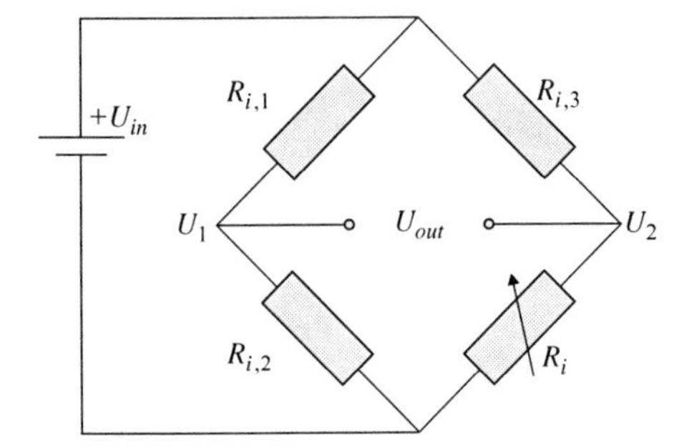

Fig. 5.8 The Wheatstone bridge

By differentiating the equation (5.10) with respect to the variable R_i, the influence of the change of the strain gauge resistance on the output voltage can be determined

$$\Delta U_{out} = -\frac{R_{i,3}U_{in}}{(R_i + R_{i,3})^2}\Delta R_i. \tag{5.11}$$

Before application, the force sensor must be calibrated, which requires the determination of a 6×6 calibration matrix transforming the six output voltages into the three forces and three torques.

5.4 Robot vision

The task of robot vision is to recognize the geometry of the robot workspace from a digital image (Figure 5.9). It is our aim to find the relation between the coordinates of a point in the two-dimensional (2D) image and the coordinates of the point in the real three-dimensional (3D) robot environment. The basic equations of optics determine the position of a point in the image plane with respect to the corresponding point in 3D space. We will therefore find the geometrical relation between the coordinates of the point $P(x_c, y_c, z_c)$ in space and the coordinates of the point $p(u, v)$ in the image.

As the aperture of the camera lenses, through which the light falls onto the image plane, is small if compared to the size of the objects manipulated by the robot, we can replace the lenses in our mathematical model by a simple pinhole. In perspective projection, points from space are projected onto the image plane by lines intersecting in a common point called the center of projection. When replacing a real camera with a pinhole camera, the center of projection is located in the center of the lenses.

When studying robot geometry and kinematics, we attached a coordinate frame to each rigid body, e.g. to robot segments or to objects manipulated by the robot. When considering robot vision, the camera itself represents a rigid body and a coordinate frame should be assigned to it. The pose of the camera will be from

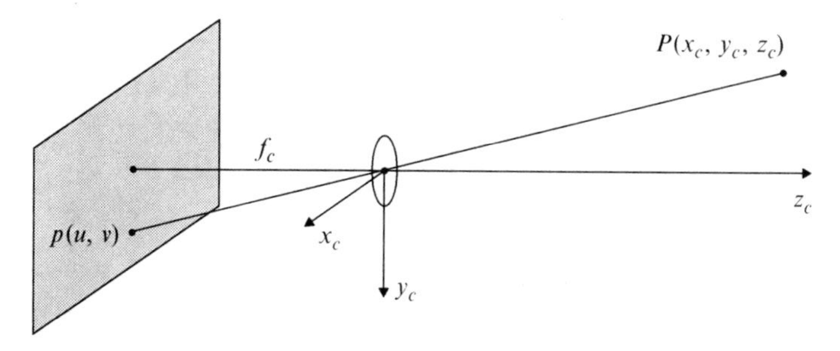

Fig. 5.9 Perspective projection

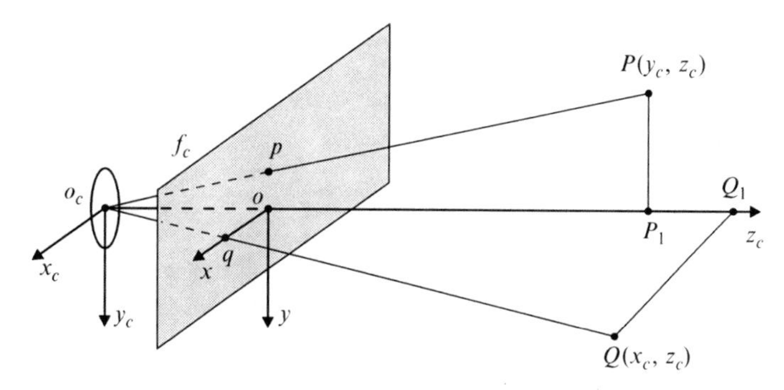

Fig. 5.10 Equivalent image plane

now on described by a corresponding coordinate frame. The z_c axis of the camera frame is directed along the optical axis, while the origin of the frame is positioned at the center of projection. We shall choose a right-handed frame where the x_c axis is parallel to the rows of the imaging sensor and the y_c axis is parallel with its columns.

The image plane is in the camera, which is placed behind the center of projection. The distance f_c between the image and the center of projection is called the focal length. In the camera frame the focal length has a negative value, as the image plane intercepts the negative z_c axis. It is more convenient to use the equivalent image plane placed at a positive z_c value (Figure 5.10). The equivalent image plane and the real image plane are symmetrical with respect to the origin of the camera frame. The geometrical properties of the objects are equivalent in both planes, and differ only in the sign.

From now on we shall call the equivalent image plane simply the image plane. Also the image plane can be considered as a rigid body to which a coordinate frame should be attached. The origin of this frame is placed in the intersection of the optical axis with the image plane. The x and y axes are parallel to the x_c and y_c axes of the camera frame.

In this way the camera has two coordinate frames, the camera frame and the image frame. Let the point P be expressed in the camera frame, while the point

p represents its projection onto the image plane. It is our aim to find the relations between the coordinates of the point P and the coordinates of its image p.

Let us first assume that the point P is located in the y_c, z_c plane of the camera frame. Its coordinates are

$$P = \begin{bmatrix} 0 \\ y_c \\ z_c \end{bmatrix}. \tag{5.12}$$

The projected point p is in this case located in the y axis of the image plane

$$p = \begin{bmatrix} 0 \\ y \end{bmatrix}. \tag{5.13}$$

Because of similarity of the triangles PP_1O_c in poO_c we can write

$$\frac{y_c}{y} = \frac{z_c}{f_c}$$

or

$$y = f_c \frac{y_c}{z_c}. \tag{5.14}$$

Let us consider also the point Q laying in the x_c, z_c plane of the camera frame. After the perspective projection of the point Q, its image q falls onto the x axis of the image frame. Because of similar triangles QQ_1O_c in qoO_c we have

$$\frac{x_c}{x} = \frac{z_c}{f_c}$$

or

$$x = f_c \frac{x_c}{z_c}. \tag{5.15}$$

In this way we obtained the relation between the coordinates $[x_c, y_c, z_c]^T$ of the point P in the camera frame and the coordinates $[x, y]^T$ of the point p in the image plane. Equations (5.14) and (5.15) represent the mathematical description of the perspective projection from a 3D onto a 2D space. Both equations can be written in the following matrix form

$$s \begin{bmatrix} x \\ y \\ 1 \end{bmatrix} = \begin{bmatrix} f_c & 0 & 0 & 0 \\ 0 & f_c & 0 & 0 \\ 0 & 0 & 1 & 0 \end{bmatrix} \begin{bmatrix} x_c \\ y_c \\ z_c \\ 1 \end{bmatrix}. \tag{5.16}$$

In equation (5.16) s is a scaling factor, while $[x, y, 1]^T$ are the coordinates of the projected point in the image frame and $[x_c, y_c, z_c, 1]^T$ are the coordinates of the original point in the camera frame.

From the matrix equation (5.16) it is not difficult to realize that we can uniquely determine the coordinates $[x, y]^T$ and the scaling factor s when knowing $[x_c, y_c, z_c]^T$. On the contrary, we cannot calculate the coordinates $[x_c, y_c, z_c]^T$ in the camera frame when only the coordinates $[x, y]^T$ in the image frame are known, but not the scaling

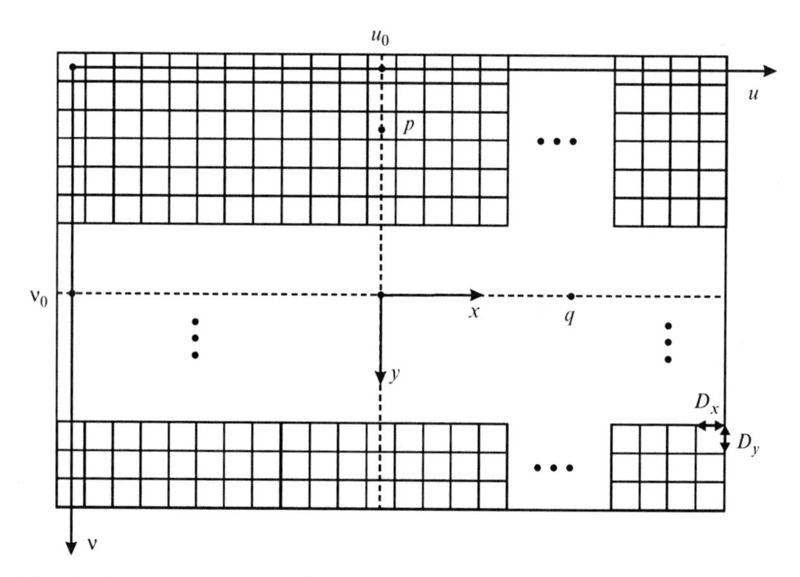

Fig. 5.11 The image plane and the index coordinate frame

factor. Equation (5.16) represents the forward projective mapping in robot vision. The calculation of $[x_c, y_c, z_c]^T$ from $[x, y]^T$ is called inverse projective mapping. When using a single camera and when having no a priori information about the size of the objects in the robot environment, a unique solution of the inverse problem cannot be found.

For the ease of programming it is more convenient to use indices, marking the position of a pixel (i.e. the smallest element of a digital image) in a 2D image instead of metric units along the x and y axes of the image frame. We shall use two indices which we shall call index coordinates of a pixel (Figure 5.11). These are the row index and the column index. In the memory, storing the digital image, the row index runs from the top of the image to its bottom, while the column index starts at the left and stops at the right edge of the image. We shall use the u axis for the column indices and the v axis for the row indices. In this way the index coordinate frame u, v belongs to each particular image. The upper left pixel is denoted either by $(0,0)$ or $(1,1)$. The index coordinates have no measuring units.

In the further text we shall find the relation between the image coordinates $[x, y]^T$ and the index coordinates $[u, v]^T$. Let us assume that the digital image was obtained as a direct output from the image sensor (A/D conversion was performed at the output of the image sensor). In this case each pixel corresponds to a particular element of the image sensor. We shall assume that the area of the image sensor is rectangular.

The origin of the image frame is in the point (u_0, v_0) of the index frame. The size of a pixel is represented by the pair (D_x, D_y). The relation between the image frame x, y and the index frame u, v is described by the following two equations:

$$\frac{x}{D_x} = u - u_0$$
$$\frac{y}{D_y} = v - v_0. \tag{5.17}$$

Equations (5.17) can be rewritten as

$$u = u_0 + \frac{x}{D_x}$$
$$v = v_0 + \frac{y}{D_y}. \tag{5.18}$$

In equations (5.18) $\frac{x}{D_x}$ and $\frac{y}{D_y}$ represent the number of digital conversions along the row and column respectively. Equations (5.18) can be rewritten in the following matrix form

$$\begin{bmatrix} u \\ v \\ 1 \end{bmatrix} = \begin{bmatrix} \frac{1}{D_x} & 0 & u_0 \\ 0 & \frac{1}{D_y} & v_0 \\ 0 & 0 & 1 \end{bmatrix} \begin{bmatrix} x \\ y \\ 1 \end{bmatrix}. \tag{5.19}$$

Using the pinhole camera model, we can now combine equations (5.16), relating the image coordinates to the camera coordinates, and equations (5.19), describing the relation between the image and index coordinates

$$s \begin{bmatrix} u \\ v \\ 1 \end{bmatrix} = \begin{bmatrix} \frac{1}{D_x} & 0 & u_0 \\ 0 & \frac{1}{D_y} & v_0 \\ 0 & 0 & 1 \end{bmatrix} \begin{bmatrix} f_c & 0 & 0 & 0 \\ 0 & f_c & 0 & 0 \\ 0 & 0 & 1 & 0 \end{bmatrix} \begin{bmatrix} x_c \\ y_c \\ z_c \\ 1 \end{bmatrix} = \begin{bmatrix} \frac{f_c}{D_x} & 0 & u_0 & 0 \\ 0 & \frac{f_c}{D_y} & v_0 & 0 \\ 0 & 0 & 1 & 0 \end{bmatrix} \begin{bmatrix} x_c \\ y_c \\ z_c \\ 1 \end{bmatrix}. \tag{5.20}$$

The above matrix can be written also in the following form

$$\mathbf{P} = \begin{bmatrix} f_x & 0 & u_0 & 0 \\ 0 & f_y & v_0 & 0 \\ 0 & 0 & 1 & 0 \end{bmatrix}. \tag{5.21}$$

The **P** matrix represents the perspective projection from the camera frame into the corresponding index coordinate frame. The variables

$$f_x = \frac{f_c}{D_x}$$
$$f_y = \frac{f_c}{D_y} \tag{5.22}$$

are the focal lengths of the camera along the x and y axes. The parameters f_x, f_y, u_0, and v_0 are called the intrinsic parameters of a camera.

In general the intrinsic parameters of the camera are not known. The specifications of the camera and the lenses are not sufficiently accurate. The intrinsic parameters of the camera are therefore obtained through the camera calibration process. When knowing the intrinsic parameters of the camera we can uniquely calculate the index coordinates $[u,v]^T$ from the given coordinates $[x_c,y_c,z_c]^T$. The coordinates $[x_c,y_c,z_c]^T$ cannot be determined from the known $[u,v]^T$ coordinates without knowing the scaling factor.

The digital image is represented by a matrix of pixels. As the index coordinates $[u,v]^T$ do not have measuring units, this means that characteristic features of the image are described more qualitatively than quantitatively. If we wish to express the distances in metric units, we must know the relation between the index coordinates $[u,v]^T$ and the coordinates $[x_r,y_r,z_r]^T$ in the 3D reference fame. Without knowing the real dimensions or the geometry of the scene it is impossible to recognize the features of the image.

Let us assume that we have a robot vision system with a single camera. The system has the image of the robot workspace as the input and is required to reproduce geometrical measurements as its output. The necessary transformations between the coordinate frames are evident from Figure 5.12.

Let us now suppose that we are in a position to recognize the point q in the image. It is our aim to determine the coordinates of the real point Q, from the coordinates of its image q. This is the problem of inverse projective mapping. In order to be able to solve the problem, we must know how the coordinates of the point q are related to the coordinates of the real point Q in the reference frame, which is the problem of forward projective mapping.

Let us solve first the problem of forward projective mapping. The point Q is given by the coordinates (x_r,y_r,z_r) in the reference coordinate frame. We wish to determine the coordinates of its image $q[u,v]$ expressed in the index frame. The frame x_c, y_c, z_c is attached to the camera. The matrix $\mathbf{M}$ represents the transformation from the reference into the camera frame

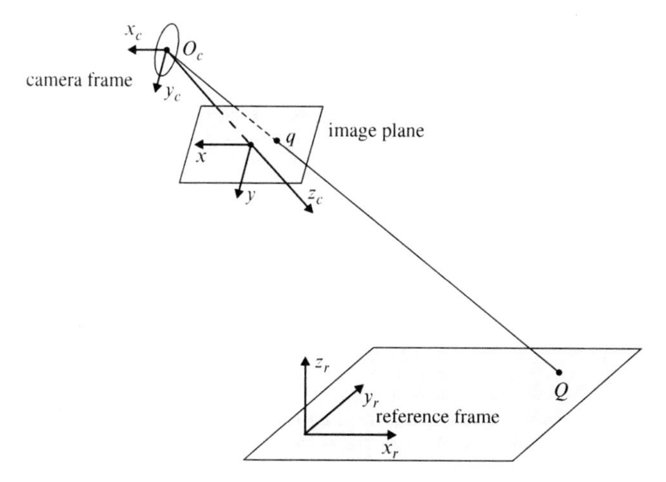

Fig. 5.12 The coordinate frames in a robot vision system

$$\begin{bmatrix} x_c \\ y_c \\ z_c \\ 1 \end{bmatrix} = \mathbf{M} \begin{bmatrix} x_r \\ y_r \\ z_r \\ 1 \end{bmatrix}. \tag{5.23}$$

By combining equations (5.23) and (5.20) we obtain

$$s \begin{bmatrix} u \\ v \\ 1 \end{bmatrix} = \mathbf{PM} \begin{bmatrix} x_r \\ y_r \\ z_r \\ 1 \end{bmatrix}. \tag{5.24}$$

The relation (5.24) describes the forward projective mapping. The elements of the **P** matrix are the intrinsic parameters of the camera, while the elements of the **M** matrix represent its extrinsic parameters. The 3×4 matrix

$$\mathbf{H} = \mathbf{PM} \tag{5.25}$$

is called the calibration matrix of the camera. It is used in the calibration process in order to determine both the intrinsic and extrinsic parameters of the camera.

In the further text we shall consider inverse projective mapping. It is our aim to determine the coordinates (x_r, y_r, z_r) of the real point Q from the known coordinates of the image point (u, v) and the calibration matrix **H**. The scaling factor s is not known. In (5.24) we have four unknowns s, x_r, y_r and z_r and only three equations for a single point in space.

Let us try with three points A, B and C (Figure 5.13). We know the distances between these three points. Their coordinates in the reference frame are

$$\{(x_{ri}, y_{ri}, z_{ri}), \quad i = 1, 2, 3\}.$$

The coordinates of the corresponding image points are

$$\{(u_i, v_i), \quad i = 1, 2, 3\}.$$

The forward projective mapping can be written in the following form

$$s_i \begin{bmatrix} u_i \\ v_i \\ 1 \end{bmatrix} = \mathbf{H} \begin{bmatrix} x_{ri} \\ y_{ri} \\ z_{ri} \\ 1 \end{bmatrix}. \tag{5.26}$$

In equation (5.26) we have 12 unknowns and 9 equations. To solve the problem we need additional three equations. These equations can be obtained from the size of the triangle represented by the points A, B and C. We shall denote the triangle sides AB, BC and CA as the lengths L_{12}, L_{23} and L_{31}

$$L_{12}^2 = (x_{r1} - x_{r2})^2 + (y_{r1} - y_{r2})^2 + (z_{r1} - z_{r2})^2$$

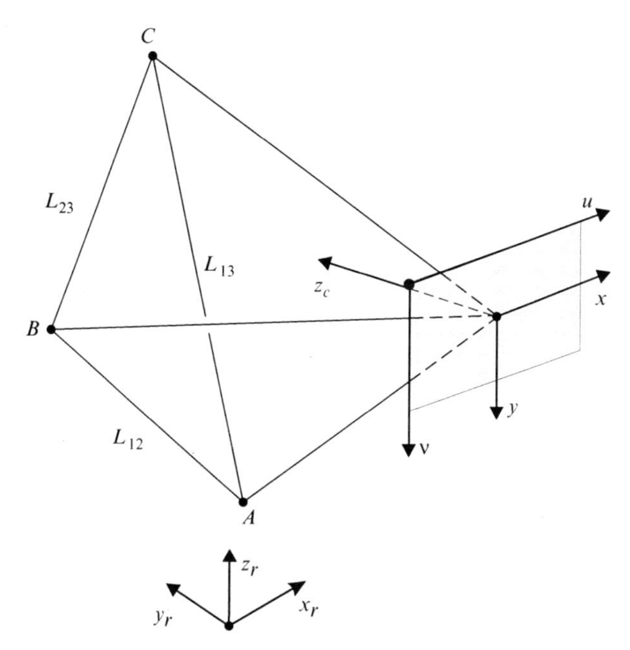

Fig. 5.13 Projection of three points in the space

$$
\begin{aligned}
L_{23}^2 &= (x_{r2} - x_{r3})^2 + (y_{r2} - y_{r3})^2 + (z_{r2} - z_{r3})^2 \\
L_{31}^2 &= (x_{r3} - x_{r1})^2 + (y_{r3} - y_{r1})^2 + (z_{r3} - z_{r1})^2 .
\end{aligned}
\tag{5.27}
$$

Now we have 12 equations for the 12 unknowns. Thus, the solution of the inverse problem exists. It is inconvenient that the last three equations are nonlinear what requires a computer for numerical solving of the equations. The approach is called model based inverse projective mapping.

Chapter 6
Trajectory planning

In previous chapters we studied mathematical models of robot mechanisms. First of all we were interested in robot kinematics and dynamics. Before applying this knowledge to robot control, we must become familiar with the planning of robot motion. The aim of trajectory planning is to generate the reference inputs to the robot control system, which will ensure that the robot end-effector will follow the desired trajectory.

Robot motion is usually defined in the rectangular world coordinate frame placed in the robot workspace most conveniently for the robot task. In the simplest task we only define the initial and the final point of the robot end-effector. The inverse kinematic model is then used to calculate the joint variables corresponding to the desired position of the robot end-effector.

6.1 Interpolation of the trajectory between two points

When moving between two points, the robot manipulator must be displaced from the initial to the final point in a given time interval t_f. In most cases we are not interested in the precise trajectory between the two points. Nevertheless, we must determine the time course of the motion for each joint variable and provide the calculated trajectory to the control input. The joint variable is either the angle ϑ for the rotational or the displacement d for the translational joint. When considering the interpolation of the trajectory we shall not distinguish between the rotational and translational joints, so that the joint variable will be more generally denoted as q. With industrial manipulators moving between two points we most often select the so called trapezoidal velocity profile. The robot movement starts at $t = 0$ with constant acceleration, followed by the phase of constant velocity and finished by the constant deceleration phase (Figure 6.1). The resulting trajectory of either the joint angle or displacement consists of the central linear interval, which is started and concluded with a parabolic segment. The initial and final velocities of the movement between the two points are zero. The duration of the constant acceleration phase is equal

T. Bajd et al., *Robotics*, Intelligent Systems, Control and Automation: Science and Engineering 43, DOI 10.1007/978-90-481-3776-3_6,

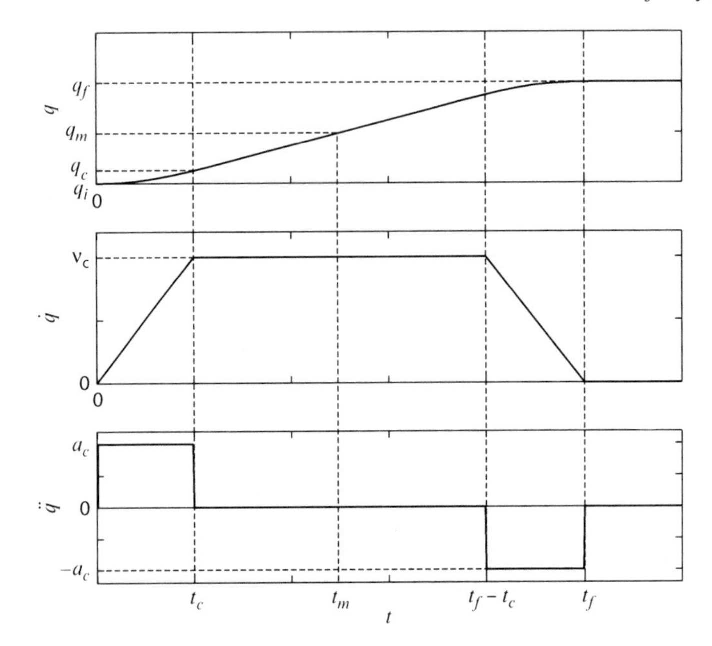

Fig. 6.1 The time dependence of the joint variables with trapezoidal velocity profile

to the interval with the constant deceleration. In both phases the magnitude of the acceleration is a_c. In this way we deal with a symmetric trajectory, where

$$q_m = \frac{q_f + q_i}{2} \quad \text{at the moment} \quad t_m = \frac{t_f}{2}. \tag{6.1}$$

The trajectory $q(t)$ must satisfy several constraints in order that the robot joint will move from the initial point q_i to the final point q_f in the required time interval t_f. The velocity at the end of the initial parabolic phase must be equal to the constant velocity in the linear phase. The velocity in the first phase is obtained from the equation describing the constant acceleration motion

$$v = a_c t. \tag{6.2}$$

At the end of the first phase we have

$$v_c = a_c t_c. \tag{6.3}$$

The velocity in the second phase can be determined by the help of Figure 6.1

$$v_c = \frac{q_m - q_c}{t_m - t_c}, \tag{6.4}$$

where q_c represents the value of the joint variable at the end of the initial parabolic phase, i.e. at the time t_c. Until that time the motion with constant acceleration a_c

takes place, so the velocity is determined by equation (6.2). The time dependence of the joint position is obtained by integrating equation (6.2)

$$q = \int v dt = a_c \int t dt = a_c \frac{t^2}{2} + q_i, \tag{6.5}$$

where the initial joint position q_i is taken as the integration constant. At the end of the first phase we have

$$q_c = q_i + \frac{1}{2} a_c t_c^2. \tag{6.6}$$

The velocity at the end of the first phase (6.3) is equal to the constant velocity in the second phase (6.4)

$$a_c t_c = \frac{q_m - q_c}{t_m - t_c}. \tag{6.7}$$

By inserting equation (6.6) into equation (6.7) and considering the expression (6.1), we obtain, after rearrangement, the following quadratic equation

$$a_c t_c^2 - a_c t_f t_c + q_f - q_i = 0. \tag{6.8}$$

The acceleration a_c is determined by the selected actuator and the dynamic properties of the robot mechanism. For chosen q_i, q_f and t_f the time interval t_c is

$$t_c = \frac{t_f}{2} - \frac{1}{2} \sqrt{\frac{t_f^2 a_c - 4(q_f - q_i)}{a_c}}. \tag{6.9}$$

To generate the movement between the initial q_i and the final position q_f the following polynomial must be generated in the first phase

$$q_1(t) = q_i + \frac{1}{2} a_c t^2 \qquad 0 \leq t \leq t_c. \tag{6.10}$$

In the second phase a linear trajectory must be generated starting in the point (t_c, q_c) with the slope v_c

$$(q - q_c) = v_c (t - t_c). \tag{6.11}$$

After rearrangement we obtain

$$q_2(t) = q_i + a_c t_c (t - \frac{t_c}{2}) \qquad t_c < t \leq (t_f - t_c). \tag{6.12}$$

In the last phase the parabolic trajectory must be generated similar to the first phase, only that now the extreme point is in (t_f, q_f) and the curve is turned upside down

$$q_3 = q_f - \frac{1}{2} a_c (t - t_f)^2 \qquad (t_f - t_c) < t \leq t_f. \tag{6.13}$$

In this way we obtained analytically the time dependence of the angle or displacement of the rotational or translational joint moving from point to point.

6.2 Interpolation by use of via points

In some robot tasks, movements of the end-effectors, more complex than point to point motions, are necessary. In welding, for example, the curved surfaces of the objects must be followed. Such trajectories can be obtained by defining, besides the initial and the final point, also the so called via points through which the robot end-effector must move.

In this chapter we shall analyze the problem, where we wish to interpolate the trajectory through n via points $\{q_1, \ldots, q_n\}$ which must be reached by the robot in time intervals $\{t_1, \ldots, t_n\}$. The interpolation will be performed with the help of trapezoidal velocity profiles. The trajectory will consist of a sequence of linear segments describing the movements between two via points and parabolic segments representing the transitions through the via points. In order to avoid the discontinuity of the first derivative at the moment t_k, the trajectory $q(t)$ must have a parabolic course in the vicinity of q_k. By doing so the second derivative in the point q_k (acceleration) remains discontinuous.

The interpolated trajectory, defined as a sequence of linear functions with parabolic transitions through the via points (the transition time Δt_k), is analytically described by the following constraints

$$q(t) = \begin{cases} a_{1,k} \cdot (t - t_k) + a_{0,k} & t_k + \frac{\Delta t_k}{2} \le t < t_{k+1} - \frac{\Delta t_{k+1}}{2} \\ b_{2,k} \cdot (t - t_k)^2 + b_{1,k} \cdot (t - t_k) + b_{0,k} & t_k - \frac{\Delta t_k}{2} \le t < t_k + \frac{\Delta t_k}{2} \end{cases} . \tag{6.14}$$

The coefficients $a_{0,k}$ and $a_{1,k}$ determine the linear parts of the trajectory, where k represents the index of the corresponding linear segment. The coefficients $b_{0,k}$, $b_{1,k}$ and $b_{2,k}$ belong to the parabolic transitions. The index k represents the consecutive number of a parabolic segment.

First, the velocities in the linear segments will be calculated from the given positions and the corresponding time intervals, as shown in Figure 6.2. We assume that the initial and final velocities are equal to zero. In this case we have

$$\dot{q}_{k-1,k} = \begin{cases} 0 & k = 1 \\ \frac{q_k - q_{k-1}}{t_k - t_{k-1}} & k = 2, \ldots, n \\ 0 & k = n + 1 . \end{cases} \tag{6.15}$$

Further, we must determine the coefficients of the linear segments $a_{0,k}$ and $a_{1,k}$. The coefficient $a_{0,k}$ can be found from the linear function (6.14), by taking into account the known position at the moment t_k, when the robot segment approaches the point q_k

$$q(t_k) = q_k = a_{1,k} \cdot (t_k - t_k) + a_{0,k} = a_{0,k}, \tag{6.16}$$

therefore

$$t = t_k \quad \Rightarrow \quad a_{0,k} = q_k \qquad k = 1, \ldots, n - 1 . \tag{6.17}$$

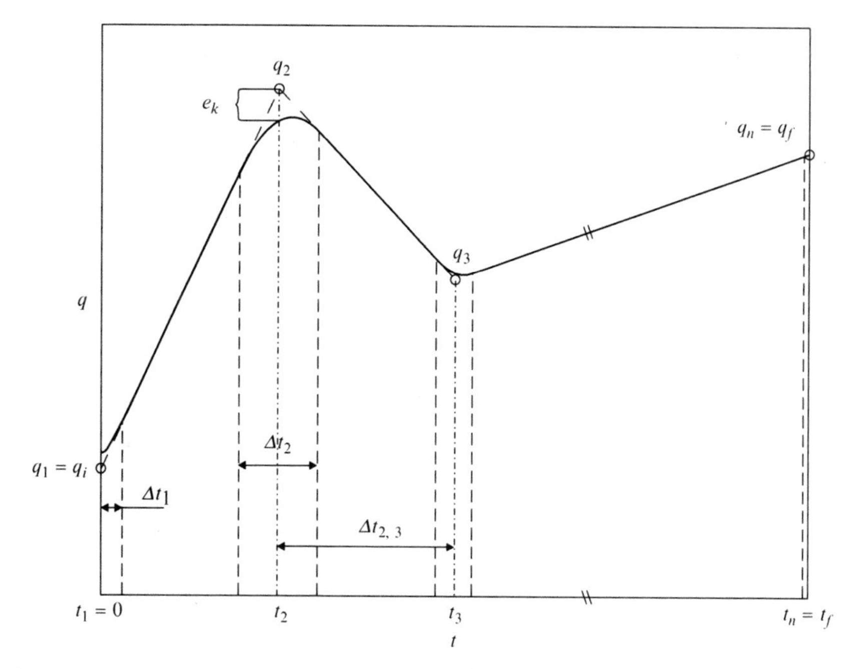

Fig. 6.2 Trajectory interpolation through n via points – linear segments with parabolic transitions are used

The coefficient $a_{1,k}$ can be determined from the time derivative of the linear function (6.14)

$$\dot{q}(t) = a_{1,k}. \tag{6.18}$$

By considering the given velocities in the time interval $t_{k,k+1}$, we obtain

$$a_{1,k} = \dot{q}_{k,k+1} \qquad k = 1,\ldots,n-1. \tag{6.19}$$

In this way the coefficients of the linear segments of the trajectory are determined and we can proceed with the coefficients of the parabolic functions. We shall assume that the transition time is predetermined as Δt_k. If the transition time is not prescribed, the absolute value of the acceleration $|\ddot{q}_k|$ in the via point must be first determined and then the transition time is calculated from the accelerations and velocities before and after the via point. In this case only the sign of the acceleration must be determined by considering the sign of the velocity difference in the via point

$$\ddot{q}_k = sign(\dot{q}_{k,k+1} - \dot{q}_{k-1,k})|\ddot{q}_k|, \tag{6.20}$$

where $sign(\cdot)$ means the sign of the expression in the brackets. Given the values of the accelerations in the via points and the velocities before and after the via point, the time of motion through the via point Δt_k is calculated (deceleration and acceleration)

$$\Delta t_k = \frac{\dot{q}_{k,k+1} - \dot{q}_{k-1,k}}{\ddot{q}_k}. \tag{6.21}$$

We shall proceed by calculating the coefficients of the quadratic functions. The required continuity of the velocity during the transition from the linear into the parabolic trajectory segment at the instant $(t_k - \frac{\Delta t_k}{2})$ and the required velocity continuity during the transition from the parabolic into the linear segment at $(t_k + \frac{\Delta t_k}{2})$ represents the starting point for the calculation of the coefficients $b_{1,k}$ and $b_{2,k}$. First, we calculate the time derivative of the quadratic function (6.14)

$$\dot{q}(t) = 2b_{2,k}(t - t_k) + b_{1,k}. \tag{6.22}$$

Assuming that the velocity at the instant $(t_k - \frac{\Delta t_k}{2})$ is $\dot{q}_{k-1,k}$, while at $(t_k + \frac{\Delta t_k}{2})$ it is $\dot{q}_{k,k+1}$, we can write

$$\begin{aligned} \dot{q}_{k-1,k} &= 2b_{2,k}\left(t_k - \frac{\Delta t_k}{2} - t_k\right) + b_{1,k} = -b_{2,k}\Delta t_k + b_{1,k} \qquad t = t_k - \frac{\Delta t_k}{2} \\ \dot{q}_{k,k+1} &= 2b_{2,k}\left(t_k + \frac{\Delta t_k}{2} - t_k\right) + b_{1,k} = b_{2,k}\Delta t_k + b_{1,k} \qquad t = t_k + \frac{\Delta t_k}{2}. \end{aligned} \tag{6.23}$$

by adding equations (6.23), the coefficient $b_{1,k}$ can be determined

$$b_{1,k} = \frac{\dot{q}_{k,k+1} + \dot{q}_{k-1,k}}{2} \qquad k = 1, \ldots, n, \tag{6.24}$$

and by subtracting equations (6.23), the coefficient $b_{2,k}$ is calculated

$$b_{2,k} = \frac{\dot{q}_{k,k+1} - \dot{q}_{k-1,k}}{2\Delta t_k} = \frac{\ddot{q}_k}{2} \qquad k = 1, \ldots, n. \tag{6.25}$$

By taking into account the continuity of the position at the instant $(t_k + \frac{\Delta t_k}{2})$, the coefficient $b_{0,k}$ of the quadratic polynomial can be calculated. At $(t_k + \frac{\Delta t_k}{2})$ the position $q(t)$, calculated from the linear function

$$q\left(t_k + \frac{\Delta t_k}{2}\right) = a_{1,k}\left(t_k + \frac{\Delta t_k}{2} - t_k\right) + a_{0,k} = \dot{q}_{k,k+1}\frac{\Delta t_k}{2} + q_k \tag{6.26}$$

equals the position $q(t)$ calculated from the quadratic function

$$\begin{aligned} q\left(t_k + \frac{\Delta t_k}{2}\right) &= b_{2,k}\left(t_k + \frac{\Delta t_k}{2} - t_k\right)^2 + b_{1,k}\left(t_k + \frac{\Delta t_k}{2} - t_k\right) + b_{0,k} \\ &= \frac{\dot{q}_{k,k+1} - \dot{q}_{k-1,k}}{2\Delta t_k}\left(\frac{\Delta t_k}{2}\right)^2 + \frac{\dot{q}_{k,k+1} + \dot{q}_{k-1,k}}{2} \cdot \frac{\Delta t_k}{2} + b_{0,k}. \end{aligned} \tag{6.27}$$

By equating (6.26) and (6.27), the coefficient $b_{0,k}$ is determined

$$b_{0,k} = q_k + (\dot{q}_{k,k+1} - \dot{q}_{k-1,k})\frac{\Delta t_k}{8}. \tag{6.28}$$

It can be verified that the calculated coefficient $b_{0,k}$ ensures also continuity of position at the instant $(t_k - \frac{\Delta t_k}{2})$. Such a choice of the coefficient $b_{0,k}$ prevents the joint trajectory to go through the point q_k. The robot only more or less approaches this point. The distance of the calculated trajectory from the reference point depends mainly on the decelerating and accelerating time interval Δt_k, which is predetermined by the required acceleration $|\ddot{q}_k|$. The error e_k of the calculated trajectory can be estimated by comparing the desired position q_k with the actual position $q(t)$ at the instant t_k, which is obtained by inserting t_k into the quadratic function (6.14)

$$e_k = q_k - q(t_k) = q_k - b_{0,k} = -(\dot{q}_{k,k+1} - \dot{q}_{k-1,k})\frac{\Delta t_k}{8}. \tag{6.29}$$

It can be noticed that the error e_k equals zero only when the velocities of the linear segments before and after the via points are equal or when the time interval Δt_k is zero meaning infinite acceleration which in reality is not possible.

The described approach to the trajectory interpolation has a minor deficiency. From equation (6.29) it can be observed that, instead of reaching the via point, the robot goes around it. As the initial and final trajectory points are also considered as via points, an error is introduced into the trajectory planning. At the starting point of the trajectory, the actual and the desired position differ by the error e_1 (Figure 6.3, the light curve shows the trajectory without correction) arising from equation (6.29). The error represents a step in the position signal which is not desired in robotics. To avoid this abrupt change in position, the first and the last trajectory point must be handled separately from the via points.

The required velocities in the starting and the final point should be zero. The velocity at the end of the time interval Δt_1 must be equal to the velocity in the first linear segment. First, we calculate the velocity in the linear part

$$\dot{q}_{1,2} = \frac{q_2 - q_1}{t_2 - t_1 - \frac{1}{2}\Delta t_1}. \tag{6.30}$$

Equation (6.30) is similar to equation (6.15) only that now $\frac{1}{2}\Delta t_1$ is subtracted in the denominator, as in the short time interval (the beginning of the parabolic segment in Figure 6.3) the position of the robot changes only to a very small extent. By doing so, higher velocity in the linear segment of the trajectory is obtained. At the end of the acceleration interval Δt_1 we have

$$\frac{q_2 - q_1}{t_2 - t_1 - \frac{1}{2}\Delta t_1} = \ddot{q}_1 \Delta t_1 \tag{6.31}$$

We must determine also the acceleration $\ddot{q}_1$ at the starting point of the trajectory. Assuming that its absolute value $|\ddot{q}_1|$ was predetermined, only the sign must be adequately selected. The choice of the sign will be performed on the basis of the positional difference (in principle the velocity difference should be taken into account

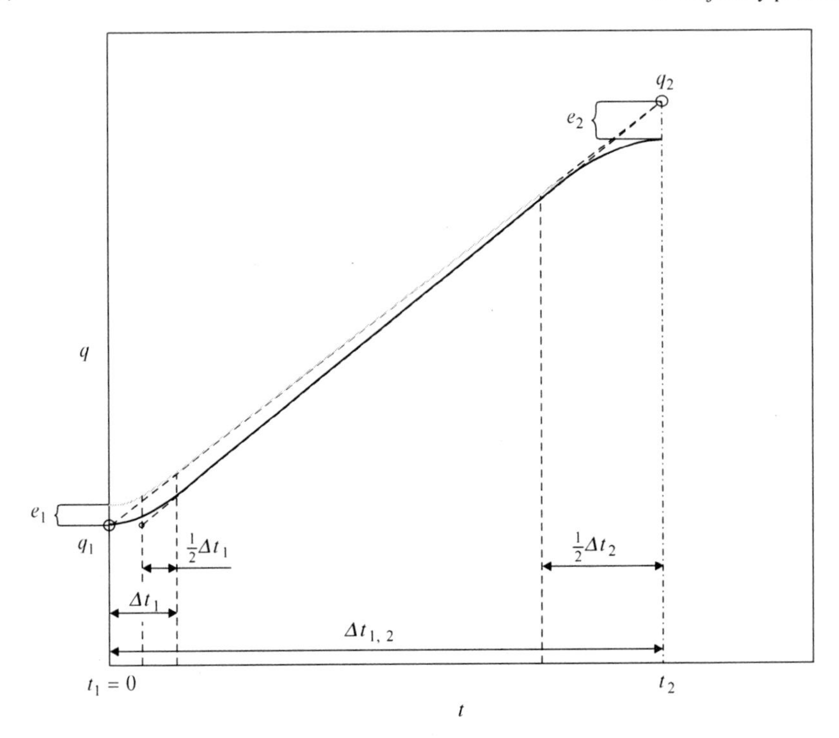

Fig. 6.3 Trajectory interpolation – enlarged presentation of the first segment of the trajectory shown in Figure 6.2. The lighter curve represents the trajectory without correction, while the darker curve shows the corrected trajectory

when determining the sign of acceleration, however the initial velocity is zero and the sign can therefore depend on the difference in positions)

$$\ddot{q}_1 = sign(q_2 - q_1)|\ddot{q}_1|. \tag{6.32}$$

From equation (6.31) the time interval Δt_1 is calculated

$$(q_2 - q_1) = \ddot{q}_1 \Delta t_1 \left(t_2 - t_1 - \frac{1}{2}\Delta t_1 \right). \tag{6.33}$$

After rearrangement we obtain

$$-\frac{1}{2}\ddot{q}_1 \Delta t_1^2 + \ddot{q}_1 (t_2 - t_1)\Delta t_1 - (q_2 - q_1) = 0 \tag{6.34}$$

so the time interval Δt_1 is

$$\Delta t_1 = \frac{-\ddot{q}_1 (t_2 - t_1) \pm \sqrt{\ddot{q}_1^2 (t_2 - t_1)^2 - 2\ddot{q}_1 (q_2 - q_1)}}{-\ddot{q}_1}, \tag{6.35}$$

and after simplifying equation (6.35)

$$\Delta t_1 = (t_2 - t_1) - \sqrt{(t_2 - t_1)^2 - \frac{2(q_2 - q_1)}{\ddot{q}_1}}. \tag{6.36}$$

In equation (6.36) the minus sign was selected for the square root, because the time interval Δt_1 must be shorter than $(t_2 - t_1)$. From equation (6.30) the velocity in the linear part of the trajectory can be calculated. As is evident from Figure 6.3 (the darker curve represents the corrected trajectory), the introduced correction eliminates the error in the initial position.

Similarly as for the first segment, the correction must be calculated also for the last segment between points q_{n-1} and q_n. The velocity in the last linear segment is

$$\dot{q}_{n-1,n} = \frac{q_n - q_{n-1}}{t_n - t_{n-1} - \frac{1}{2}\Delta t_n}. \tag{6.37}$$

In the denominator of equation (6.37) the value $\frac{1}{2}\Delta t_n$ was subtracted, as immediately before the complete stop of the robot, its position changes only very little. At the transition from the last linear segment into the last parabolic segment the velocities are equal

$$\frac{q_n - q_{n-1}}{t_n - t_{n-1} - \frac{1}{2}\Delta t_n} = \ddot{q}_n \Delta t_n. \tag{6.38}$$

The acceleration (deceleration) of the last parabolic segment is determined on the basis of the positional difference

$$\ddot{q}_n = sign(q_{n-1} - q_n)|\ddot{q}_n|. \tag{6.39}$$

By inserting the above equation into equation (6.38), we calculate, in a similar way as for the first parabolic segment, also the duration of the last parabolic segment

$$\Delta t_n = (t_n - t_{n-1}) - \sqrt{(t_n - t_{n-1})^2 - \frac{2(q_n - q_{n-1})}{\ddot{q}_n}}. \tag{6.40}$$

From equation (6.37) the velocity of the last linear segment can be determined. By considering the corrections at the start and at the end of the trajectory, the time course through the via points is calculated. In this way the entire trajectory was interpolated at the n points.

Chapter 7
Robot control

The problem of robot control can be explained as a computation of the forces or torques which must be generated by the actuators in order to successfully accomplish the robot task. The appropriate working conditions must be ensured both during the transient period as well as in the stationary state. The robot task can be presented either as the execution of the motions in a free space, where position control is performed, or in contact with the environment, where control of the contact force is required. First, we shall study the position control of a robot mechanism which is not in contact with its environment. Then, in the further text we shall upgrade the position control with the force control.

The problem of robot control is not unique. There exist various methods which differ in their complexity and in the effectiveness of robot actions. The choice of the control method depends on the robot task. An important difference is, for example, between the task where the robot end-effector must accurately follow the prescribed trajectory (e.g. laser welding) and another task where it is only required that the robot end-effector reaches the desired final pose, while the details of the trajectory between the initial and the final point are not important (e.g. palletizing). The mechanical structure of the robot mechanism also influences the selection of the appropriate control method. The control of a cartesian robot manipulator in general differs from the control of an anthropomorphic robot.

Robot control usually takes place in the world coordinate frame, which is defined by the user and is called also the coordinate frame of the robot task. Instead of world coordinate frame we often use a shorter expression, namely external coordinates. We are predominantly interested in the pose of the robot end-effector expressed in the external coordinates and rarely in the joint positions, which are also called internal coordinates. Nevertheless, we must be aware that in all cases we directly control the internal coordinates i.e. joint angles or displacements. The end-effector pose is only controlled indirectly. It is determined by the kinematic model of the robot mechanism and the given values of the internal coordinates.

Figure 7.1 shows a general robot control system. The input to the control system is the desired pose of the robot end-effector, which is obtained by using trajectory interpolation methods, introduced in the previous chapter. The variable $\mathbf{x}_r$ represents

T. Bajd et al., *Robotics*, Intelligent Systems, Control and Automation: Science and Engineering 43, DOI 10.1007/978-90-481-3776-3_7,

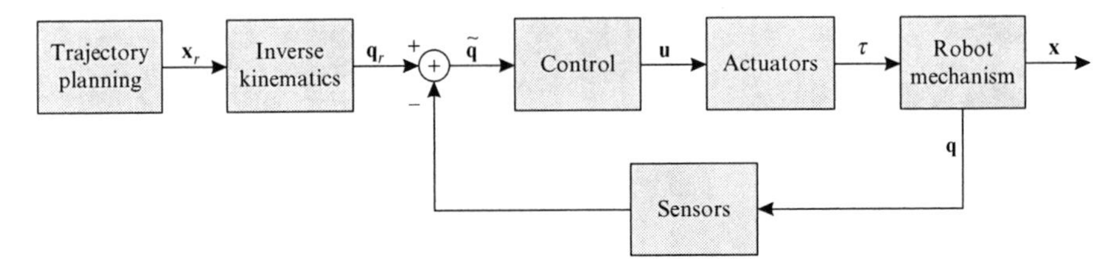

Fig. 7.1 A general robot control system

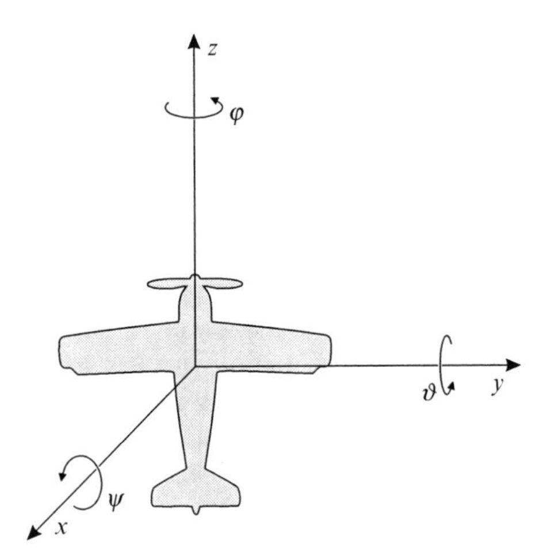

Fig. 7.2 The RPY description of the orientation

the desired, i.e. the reference pose of the robot end-effector. The $\mathbf{x}$ vector, describing the actual pose of the robot end-effector in general comprises six variables. Three of them define the position of the robot end-point, while the other three determine the orientation of the robot end-effector. Thus, we write $\mathbf{x} = \begin{bmatrix} x & y & z & \varphi & \vartheta & \psi \end{bmatrix}^T$. The position of the robot end-effector is determined by the vector from the origin of the world coordinate frame to the robot end-point. The orientation of the end-effector can be presented in various ways. One of the possible descriptions is the so called RPY notation, arising from aeronautics and shown in Figure 7.2. The orientation is determined by the angle φ around the z axis (Roll), the angle ϑ around the y axis (Pitch) and the angle ψ around the x axis (Yaw).

By the use of the inverse kinematics algorithm, the internal coordinates $\mathbf{q}_r$, corresponding to the desired end-effector pose, are calculated. The variable q_r represents the joint position, i.e. the angle ϑ for the rotational joint and the distance d for the translational joint. The desired internal coordinates are compared to the actual internal coordinates in the robot control system. On the basis of the positional error $\tilde{\mathbf{q}}$, the control system output $\mathbf{u}$ is calculated. The output $\mathbf{u}$ is converted from a digital into an analogue signal, amplified and delivered to the robot actuators. The actuators ensure the forces or torques necessary for the required robot motion. The robot motion is assessed by the sensors which were described in the chapter devoted to robot sensors.

7.1 Control of the robot in internal coordinates

The simplest robot control approach is based on controllers where the control loop is closed separately for each particular degree of freedom. Such controllers are suitable for control of independent second order systems with constant inertial and damping parameters. This approach is less suitable for robotic systems characterized by nonlinear and time varying behavior.

7.1.1 PD control of position

First, a simple proportional-derivative (PD) controller will be analyzed. The basic control scheme is shown in Figure 7.3. The control is based on calculation of the positional error and determination of control parameters, which enable reduction or suppression of the error. The positional error is reduced for each joint separately, which means that as many controllers are to be developed as there are degrees of freedom. The reference positions $\mathbf{q}_r$ are compared to the actual positions of the robot joints $\mathbf{q}$

$$\tilde{\mathbf{q}} = \mathbf{q}_r - \mathbf{q}. \tag{7.1}$$

The positional error $\tilde{\mathbf{q}}$ is amplified by the proportional position gain $\mathbf{K}_p$. As a robot manipulator has several degrees of freedom, the error $\tilde{\mathbf{q}}$ is expressed as a vector, while $\mathbf{K}_p$ is a diagonal matrix of the gains of all joint controllers. The calculated control input provokes robot motion in the direction of reduction of the positional error. As the actuation of the robot motors is proportional to the error, it can occur that the robot will overshoot instead of stopping in the desired position. Such overshoots are not allowed in robotics, as they may result in collisions with objects in the robot vicinity. To ensure safe and stable robot actions, a velocity closed loop is introduced with a negative sign. The velocity closed loop brings damping into the system. It is represented by the actual joint velocities $\dot{\mathbf{q}}$ multiplied by a diagonal matrix of velocity gains $\mathbf{K}_d$. The control law can be written in the following form

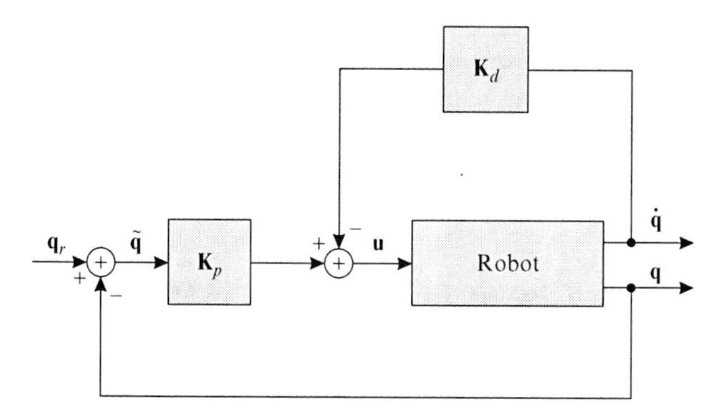

Fig. 7.3 PD position control with high damping

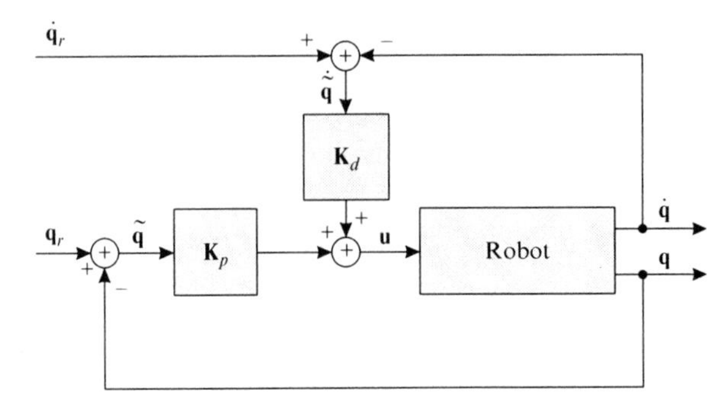

Fig. 7.4 PD position control

$$\mathbf{u} = \mathbf{K}_p(\mathbf{q}_r - \mathbf{q}) - \mathbf{K}_d\dot{\mathbf{q}}, \tag{7.2}$$

where $\mathbf{u}$ represents the control inputs, i.e. the joint forces or torques, which must be provided by the actuators. From equation (7.2) we can notice that at higher velocities of robot motions, the velocity control loop reduces the joint actuation and, by damping the system, ensures robot stability.

The control method shown in Figure 7.3 provides high damping of the system in the fastest part of the trajectory, which is usually not necessary. Such behavior of the controller can be avoided by upgrading the PD controller with the reference velocity signal. This signal is obtained as the numerical derivative of the desired position. The velocity error is used as control input

$$\dot{\tilde{\mathbf{q}}} = \dot{\mathbf{q}}_r - \dot{\mathbf{q}}. \tag{7.3}$$

The control algorithm demonstrated in Figure 7.4 can be written as

$$\mathbf{u} = \mathbf{K}_p(\mathbf{q}_r - \mathbf{q}) + \mathbf{K}_d(\dot{\mathbf{q}}_r - \dot{\mathbf{q}}). \tag{7.4}$$

As the difference between the reference velocity $\dot{\mathbf{q}}_r$ and $\dot{\mathbf{q}}$ is used instead of the total velocity $\dot{\mathbf{q}}$, the damping effect is reduced. For a positive difference the control loop can even accelerate the robot motion.

The synthesis of the PD position controller consists of determining the matrices $\mathbf{K}_p$ and $\mathbf{K}_d$. For fast response, the $\mathbf{K}_p$ gains must be high. By proper choice of the $\mathbf{K}_d$ gains, critical damping of the robot systems is obtained. The critical damping ensures fast response without overshoot. Such controllers must be built for each joint separately. The behavior of each controller is entirely independent of the controllers belonging to the other joints of the robot mechanism.

7.1.2 PD control of position with gravity compensation

In the chapter on robot dynamics we found that the robot mechanism is under the influence of inertial, Coriolis, centripetal and gravitational forces (4.46). In general also friction forces, occurring in robot joints, must be included in the robot dynamic model. In a somewhat simplified model, only viscous friction, being proportional to the joint velocity, will be taken into account ($\mathbf{F}_v$ is a diagonal matrix of the joint friction coefficients). The enumerated forces must be overcome by the robot actuators which is evident from the following equation, similar to equation (4.46)

$$\mathbf{B}(\mathbf{q})\ddot{\mathbf{q}} + \mathbf{C}(\mathbf{q},\dot{\mathbf{q}})\dot{\mathbf{q}} + \mathbf{F}_v\dot{\mathbf{q}} + \mathbf{g}(\mathbf{q}) = \tau. \tag{7.5}$$

When developing the PD controller, we did not pay attention to the specific forces influencing the robot mechanism. The robot controller calculated the required actuation forces solely on the basis of the difference between the desired and the actual joint positions. Such a controller cannot predict the force necessary to produce the desired robot motion. As the force is calculated from the positional error, this means that in general the error is never equal to zero. When knowing the dynamic robot model, we can predict the forces which are necessary for the performance of a particular robot motion. These forces are then generated by the robot motors regardless of the positional error signal.

In quasi-static conditions, when the robot is moving slowly, we can assume zero accelerations $\ddot{\mathbf{q}} \approx \mathbf{0}$ and velocities $\dot{\mathbf{q}} \approx \mathbf{0}$. The robot dynamic model is simplified as follows

$$\tau \approx \mathbf{g}(\mathbf{q}). \tag{7.6}$$

According to equation (7.6), the robot motors must above all compensate the gravity effect. The model of gravitational effects $\hat{\mathbf{g}}(\mathbf{q})$ (the circumflex denotes the robot model), which is a good approximation of the actual gravitational forces $\mathbf{g}(\mathbf{q})$, can be implemented in the control algorithm as shown in Figure 7.5. The PD controller, shown in Figure 7.3, was upgraded with an additional control loop, which calculates the gravitational forces from the actual robot position and directly adds them to the controller output. The control algorithm shown in Figure 7.5 can be written as follows

$$\mathbf{u} = \mathbf{K}_p(\mathbf{q}_r - \mathbf{q}) - \mathbf{K}_d\dot{\mathbf{q}} + \hat{\mathbf{g}}(\mathbf{q}). \tag{7.7}$$

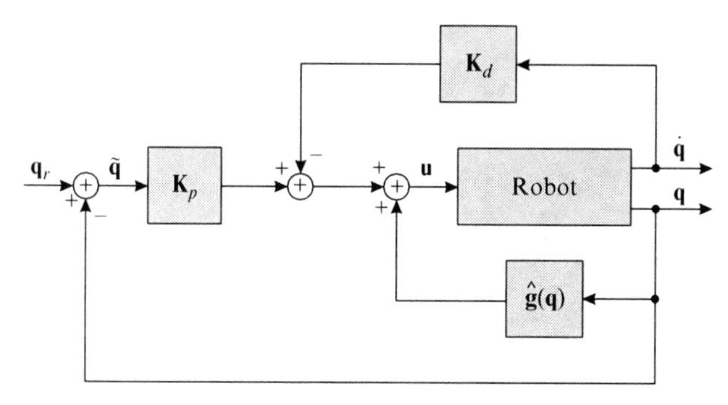

Fig. 7.5 PD control with gravity compensation

By introducing gravity compensation, the burden of reducing the errors caused by gravity, is taken away from the PD controller. In this way the errors in trajectory tracking are significantly reduced.

7.1.3 Control of the robot based on inverse dynamics

When studying the PD controller with gravity compensation, we investigated the robot dynamic model in order to improve the efficiency of the control method. With the control method based on inverse dynamics, this concept will be further upgraded. From the equations describing the dynamic behavior of a two-segment robot manipulator (4.46), we can clearly observe that the robot model is nonlinear. A linear controller, such as the PD controller, is therefore not the best choice.

We shall derive the new control scheme from the robot dynamic model described by equation (7.5). Let us assume that the torques τ, generated by the motors, are equal to the control outputs $\mathbf{u}$. Equation (7.5) can be rewritten

$$\mathbf{B}(\mathbf{q})\ddot{\mathbf{q}} + \mathbf{C}(\mathbf{q},\dot{\mathbf{q}})\dot{\mathbf{q}} + \mathbf{F}_v\dot{\mathbf{q}} + \mathbf{g}(\mathbf{q}) = \mathbf{u}. \tag{7.8}$$

In the next step we will determine the direct robot dynamic model, which describes robot motions under the influence of the given joint torques. First we express the acceleration $\ddot{\mathbf{q}}$ from equation (7.8)

$$\ddot{\mathbf{q}} = \mathbf{B}^{-1}(\mathbf{q})\left(\mathbf{u} - (\mathbf{C}(\mathbf{q},\dot{\mathbf{q}})\dot{\mathbf{q}} + \mathbf{F}_v\dot{\mathbf{q}} + \mathbf{g}(\mathbf{q}))\right). \tag{7.9}$$

By integrating the acceleration, while taking into account the initial velocity value, the velocity of robot motion is obtained. By integrating the velocity, while taking into account the initial position, we calculate the actual positions in the robot joints. The direct dynamic model of a robot mechanism is shown in Figure 7.6.

In order to simplify the dynamic equations, we shall define a new variable $\mathbf{n}(\mathbf{q},\dot{\mathbf{q}})$ comprising all dynamic components except the inertial component

$$\mathbf{n}(\mathbf{q},\dot{\mathbf{q}}) = \mathbf{C}(\mathbf{q},\dot{\mathbf{q}})\dot{\mathbf{q}} + \mathbf{F}_v\dot{\mathbf{q}} + \mathbf{g}(\mathbf{q}). \tag{7.10}$$

The robot dynamic model can be described with the following shorter equation

$$\mathbf{B}(\mathbf{q})\ddot{\mathbf{q}} + \mathbf{n}(\mathbf{q},\dot{\mathbf{q}}) = \tau. \tag{7.11}$$

In the same way also equation (7.9) can be written in a shorter form

$$\ddot{\mathbf{q}} = \mathbf{B}^{-1}(\mathbf{q})\left(\mathbf{u} - \mathbf{n}(\mathbf{q},\dot{\mathbf{q}})\right). \tag{7.12}$$

Let us assume that the robot dynamic model is known. The inertial matrix $\hat{\mathbf{B}}(\mathbf{q})$ is an approximation of the real values $\mathbf{B}(\mathbf{q})$, while $\hat{\mathbf{n}}(\mathbf{q},\dot{\mathbf{q}})$ represents an approximation of $\mathbf{n}(\mathbf{q},\dot{\mathbf{q}})$ as follows

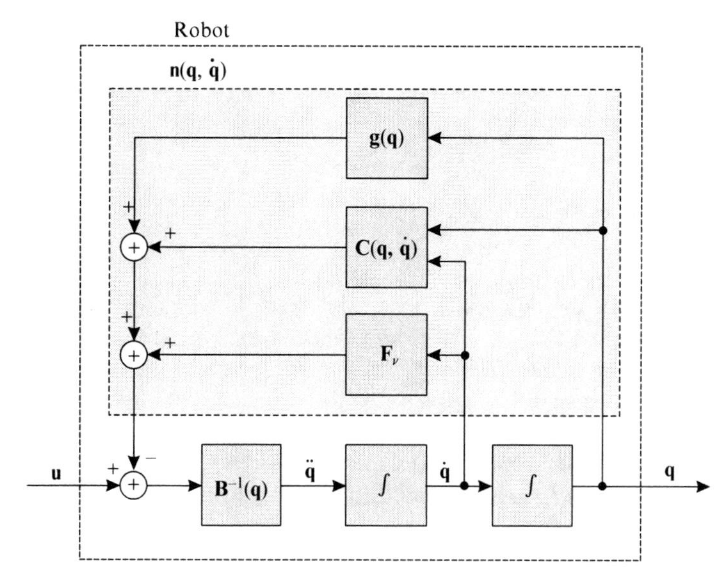

Fig. 7.6 The direct dynamic model of a robot mechanism

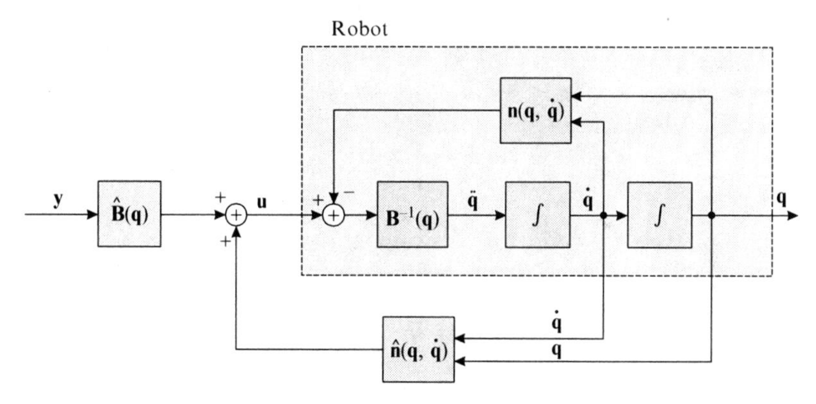

Fig. 7.7 Linearization of the control system by implementing the inverse dynamic model

$$\hat{\mathbf{n}}(\mathbf{q},\dot{\mathbf{q}}) = \hat{\mathbf{C}}(\mathbf{q},\dot{\mathbf{q}})\dot{\mathbf{q}} + \hat{\mathbf{F}}_v\dot{\mathbf{q}} + \hat{\mathbf{g}}(\mathbf{q}). \tag{7.13}$$

The controller output $\mathbf{u}$ is determined by the following equation

$$\mathbf{u} = \hat{\mathbf{B}}(\mathbf{q})\mathbf{y} + \hat{\mathbf{n}}(\mathbf{q},\dot{\mathbf{q}}), \tag{7.14}$$

where the approximate inverse dynamic model of the robot was used. The system, combining equations (7.12) and (7.14), is shown in Figure 7.7.

Let us assume the equivalence $\hat{\mathbf{B}}(\mathbf{q}) = \mathbf{B}(\mathbf{q})$ and $\hat{\mathbf{n}}(\mathbf{q},\dot{\mathbf{q}}) = \mathbf{n}(\mathbf{q},\dot{\mathbf{q}})$. In Figure 7.7 we observe that the signals $\hat{\mathbf{n}}(\mathbf{q},\dot{\mathbf{q}})$ and $\mathbf{n}(\mathbf{q},\dot{\mathbf{q}})$ subtract, as one is presented

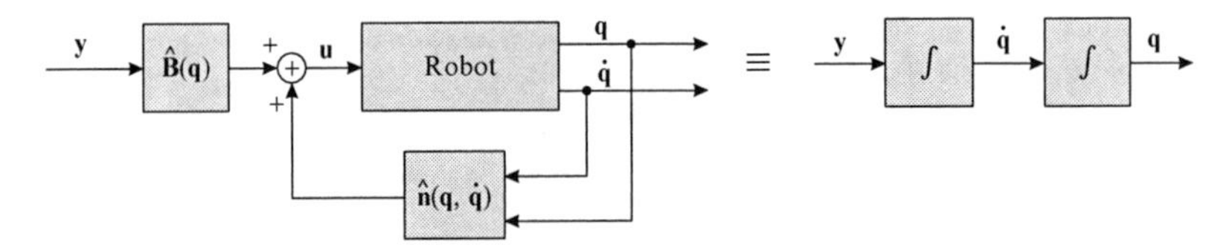

Fig. 7.8 The linearized system

with a positive and the other with a negative sign. In a similar way, the product of matrices $\hat{\mathbf{B}}(\mathbf{q})$ and $\mathbf{B}^{-1}(\mathbf{q})$ results in a unit matrix, which can be omitted. The simplified system is shown in Figure 7.8. By implementing the inverse dynamics (7.14), the control system is linearized, as there are only two integrators between the input $\mathbf{y}$ and the output $\mathbf{q}$. The system is not only linear, but is also decoupled, as e.g. the first element of the vector $\mathbf{y}$ only influences the first element of the position vector $\mathbf{q}$. From Figure 7.8 it is also not difficult to realize that the variable $\mathbf{y}$ has the characteristics of acceleration, thus

$$\mathbf{y} = \ddot{\mathbf{q}}. \tag{7.15}$$

In an ideal case, it would suffice to determine the desired joint accelerations as the second derivatives of the desired joint positions and the control system will track the prescribed joint trajectories. As we never have a fully accurate dynamic model of the robot, always a difference will occur between the desired and the actual joint positions and will increase with time. The positional error is defined by

$$\tilde{\mathbf{q}} = \mathbf{q}_r - \mathbf{q}, \tag{7.16}$$

where $\mathbf{q}_r$ represents the desired robot position. In a similar way also the velocity error can be defined as the difference between the desired and the actual velocity

$$\dot{\tilde{\mathbf{q}}} = \dot{\mathbf{q}}_r - \dot{\mathbf{q}}. \tag{7.17}$$

The vector $\mathbf{y}$, having the acceleration characteristics, can be now written as

$$\mathbf{y} = \ddot{\mathbf{q}}_r + \mathbf{K}_p(\mathbf{q}_r - \mathbf{q}) + \mathbf{K}_d(\dot{\mathbf{q}}_r - \dot{\mathbf{q}}). \tag{7.18}$$

It consists of the reference acceleration $\ddot{\mathbf{q}}_r$ and two contributing signals which depend on the errors of position and velocity. These two signals suppress the error arising because of the imperfectly modeled dynamics. The complete control scheme is shown in Figure 7.9.

By considering equation (7.18) and the equality $\mathbf{y} = \ddot{\mathbf{q}}$, the differential equation describing the robot dynamics can be written as

$$\ddot{\tilde{\mathbf{q}}} + \mathbf{K}_d\dot{\tilde{\mathbf{q}}} + \mathbf{K}_p\tilde{\mathbf{q}} = \mathbf{0}, \tag{7.19}$$

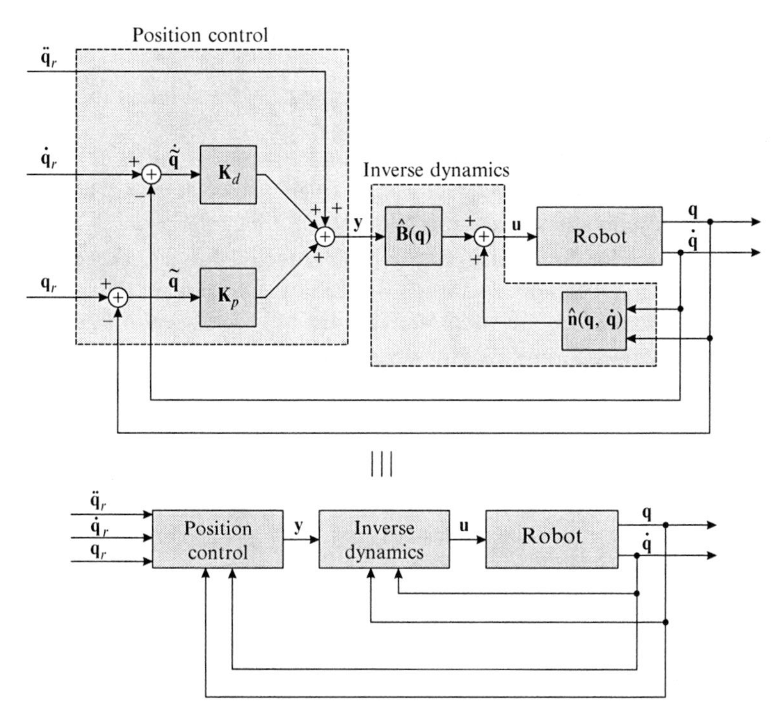

Fig. 7.9 Control of the robot based on inverse dynamics

where the acceleration error $\ddot{\tilde{\mathbf{q}}} = \ddot{\mathbf{q}}_r - \ddot{\mathbf{q}}$ was introduced. The differential equation (7.19) describes the time dependence of the control error as it approaches zero. The dynamics of the response is determined by the gains $\mathbf{K}_p$ and $\mathbf{K}_d$.

7.2 Control of the robot in external coordinates

All the control schemes studied up to now were based on control of the internal coordinates, i.e. joint positions. The desired positions, velocities and accelerations were determined by the robot joint variables. Usually we are more interested in the motion of the robot end-effector than in the displacements of particular robot joints. At the tip of the robot, different tools are attached to accomplish various robot tasks. In the further text we shall focus on the robot control in the external coordinates.

7.2.1 Control based on the transposed Jacobian matrix

The control method is based on the already known equation (4.17) connecting the forces acting at the robot end-effector with the joint torques. The relation is defined by the use of the transposed Jacobian matrix

$$\tau = \mathbf{J}^T(\mathbf{q})\mathbf{f}, \tag{7.20}$$

where the vector τ represents the joint torques and $\mathbf{f}$ is the force at the robot end-point.

It is our aim to control the pose of the robot end-effector, where its desired pose is defined by the vector $\mathbf{x}_r$ and the actual pose is given by the vector $\mathbf{x}$. The vectors $\mathbf{x}_r$ and $\mathbf{x}$ in general comprise six variables, three determining the position of the robot end-point and three for the orientation of the end-effector, thus $\mathbf{x} = \begin{bmatrix} x & y & z & \varphi & \vartheta & \psi \end{bmatrix}^T$. Robots are usually not equipped with sensors assessing the pose of the end-effector; robot sensors measure the joint variables. The pose of the robot end-effector must be therefore determined by using the equations of the direct kinematic model $\mathbf{x} = \mathbf{k}(\mathbf{q})$ introduced in the chapter on robot kinematics (4.4). The positional error of the robot end-effector is calculated as

$$\tilde{\mathbf{x}} = \mathbf{x}_r - \mathbf{x} = \mathbf{x}_r - \mathbf{k}(\mathbf{q}). \tag{7.21}$$

The positional error must be reduced to zero. A simple proportional control system with the gain matrix $\mathbf{K}_p$ is introduced

$$\mathbf{f} = \mathbf{K}_p\tilde{\mathbf{x}}. \tag{7.22}$$

When analyzing equation (7.22) more closely, we find that it reminds us of the equation describing the behavior of a spring, where the force is proportional to the spring elongation. This consideration helps us to explain the introduced control principle. Let as imagine that there are six springs virtually attached to the robot end-effector, one spring for each degree of freedom (three for position and three for orientation). When the robot moves away from the desired pose, the springs are elongated and pull the robot end-effector into the desired pose with the force proportional to the positional error. The force $\mathbf{f}$ therefore pushes the robot end-effector towards the desired pose. As the robot displacement can only be produced by the motors in the joints, the variables controlling the motors must be calculated from the force $\mathbf{f}$. This calculation is performed by the help of the transposed Jacobian matrix as shown in equation (7.20)

$$\mathbf{u} = \mathbf{J}^T(\mathbf{q})\mathbf{f}. \tag{7.23}$$

The vector $\mathbf{u}$ represents the desired joint torques. The control method based on the transposed Jacobian matrix is shown in Figure 7.10.

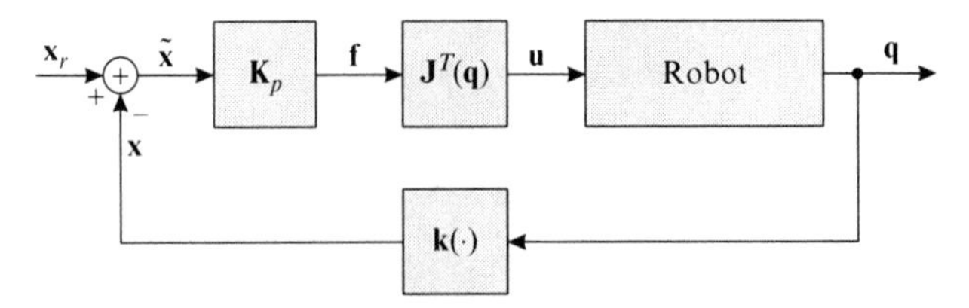

Fig. 7.10 Control based on the transposed Jacobian matrix

7.2.2 Control based on the inverse Jacobian matrix

The control method is based on the relation between the joint velocities and the velocities of the robot end-point (4.10), which is given by the Jacobian matrix. In equation (4.10) we emphasize the time derivatives of external coordinates $\mathbf{x}$ and internal coordinates $\mathbf{q}$

$$\dot{\mathbf{x}} = \mathbf{J}(\mathbf{q})\dot{\mathbf{q}} \quad \Leftrightarrow \quad \frac{d\mathbf{x}}{dt} = \mathbf{J}(\mathbf{q})\frac{d\mathbf{q}}{dt}. \tag{7.24}$$

As dt appears in the denominator on both sides of equation (7.24), it can be omitted. In this way we obtain the relation between changes of the internal coordinates and changes of the pose of the robot end-point

$$d\mathbf{x} = \mathbf{J}(\mathbf{q})d\mathbf{q}. \tag{7.25}$$

Equation (7.25) is valid only for small displacements.

As with the previously studied control method, based on the transposed Jacobian matrix, also in this case we first calculate the error of the pose of the robot end-point by using equation (7.21). When the error in the pose is small, we can calculate the positional error in the internal coordinates by the inverse relation (7.25)

$$\tilde{\mathbf{q}} = \mathbf{J}^{-1}(\mathbf{q})\tilde{\mathbf{x}}. \tag{7.26}$$

In this way the control method is translated to the known method of robot control in the internal coordinates. In the simplest example, based on the proportional controller, we can write

$$\mathbf{u} = \mathbf{K}_p\tilde{\mathbf{q}}. \tag{7.27}$$

The control method, based on the inverse Jacobian matrix, is shown in Figure 7.11.

7.2.3 PD control of position with gravity compensation

The PD control of position with gravity compensation was already studied in detail for the internal coordinates. Now we shall derive the analogue control algorithm in

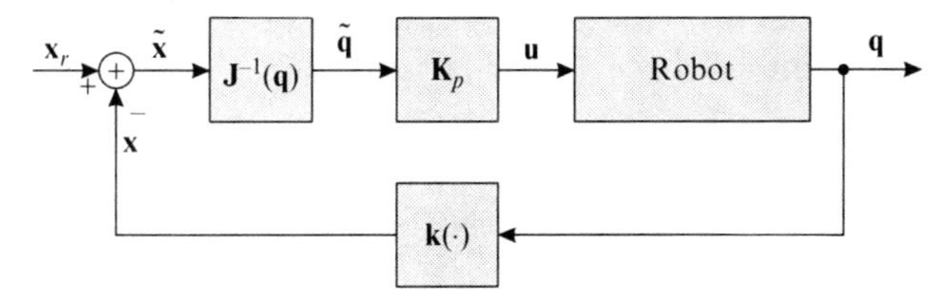

Fig. 7.11 Control based on the inverse Jacobian matrix

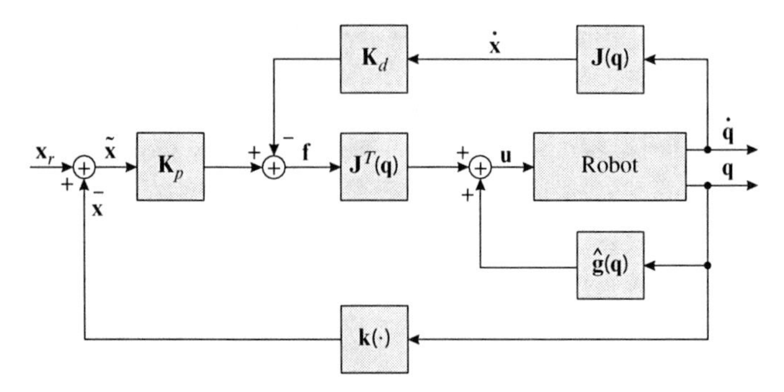

Fig. 7.12 PD control with gravity compensation in external coordinates

the external coordinates. The starting point will be equation (7.21) expressing the error of the pose of the end-effector. The velocity of the robot end-point is calculated with the help of the Jacobian matrix from the joint velocities

$$\dot{\mathbf{x}} = \mathbf{J}(\mathbf{q})\dot{\mathbf{q}}. \tag{7.28}$$

The equation describing the PD controller in external coordinates is analogous to that written in the internal coordinates (7.2)

$$\mathbf{f} = \mathbf{K}_p\tilde{\mathbf{x}} - \mathbf{K}_d\dot{\mathbf{x}}. \tag{7.29}$$

In equation (7.29), the pose error is multiplied by the matrix of the positional gains $\mathbf{K}_p$, while the velocity error is multiplied by the matrix $\mathbf{K}_d$. The negative sign of the velocity error introduces damping into the system. The joint torques are calculated from the force $\mathbf{f}$, acting at the tip of the robot, with the help of the transposed Jacobian matrix (in a similar way as in equation (7.23)) and by adding the component compensating gravity (as in equation (7.7)). The control algorithm is written as

$$\mathbf{u} = \mathbf{J}^T(\mathbf{q})\mathbf{f} + \hat{\mathbf{g}}(\mathbf{q}). \tag{7.30}$$

The complete control scheme is shown in Figure 7.12.

7.2.4 Control of the robot based on inverse dynamics

In the chapter on the control of robots in the internal coordinates, the following controller based on inverse dynamics was introduced

$$\mathbf{u} = \hat{\mathbf{B}}(\mathbf{q})\mathbf{y} + \hat{\mathbf{n}}(\mathbf{q}, \dot{\mathbf{q}}). \tag{7.31}$$

We also learned that the vector $\mathbf{y}$ has the characteristics of acceleration

$$\mathbf{y} = \ddot{\mathbf{q}}, \tag{7.32}$$

which was determined in such a way, that the robot tracked the desired trajectory expressed in the internal coordinates. As it is our aim to develop a control method in the external coordinates, the $\mathbf{y}$ signal must be adequately adapted. Equation (7.31), linearizing the system, remains unchanged.

We shall again start from the equation relating the joint velocities to the robot end-effector velocities

$$\dot{\mathbf{x}} = \mathbf{J}(\mathbf{q})\dot{\mathbf{q}}. \tag{7.33}$$

By calculating the time derivative of equation (7.33), we obtain

$$\ddot{\mathbf{x}} = \mathbf{J}(\mathbf{q})\ddot{\mathbf{q}} + \dot{\mathbf{J}}(\mathbf{q},\dot{\mathbf{q}})\dot{\mathbf{q}}. \tag{7.34}$$

The error of the pose of the robot end-effector is determined as the difference between its desired and its actual pose

$$\tilde{\mathbf{x}} = \mathbf{x}_r - \mathbf{x} = \mathbf{x}_r - \mathbf{k}(\mathbf{q}). \tag{7.35}$$

In a similar way the velocity error of the robot end-effector is determined

$$\dot{\tilde{\mathbf{x}}} = \dot{\mathbf{x}}_r - \dot{\mathbf{x}} = \dot{\mathbf{x}}_r - \mathbf{J}(\mathbf{q})\dot{\mathbf{q}}. \tag{7.36}$$

The acceleration error is the difference between the desired and the actual acceleration

$$\ddot{\tilde{\mathbf{x}}} = \ddot{\mathbf{x}}_r - \ddot{\mathbf{x}}. \tag{7.37}$$

When developing the inverse dynamics based controller in the internal coordinates, equation (7.19) was derived describing the dynamics of the control error in the form $\ddot{\tilde{\mathbf{q}}} + \mathbf{K}_d\dot{\tilde{\mathbf{q}}} + \mathbf{K}_p\tilde{\mathbf{q}} = \mathbf{0}$. An analogous equation can be written for the error of the end-effector pose. From this equation the acceleration $\ddot{\mathbf{x}}$ of the robot end-effector can be expressed

$$\ddot{\tilde{\mathbf{x}}} + \mathbf{K}_d\dot{\tilde{\mathbf{x}}} + \mathbf{K}_p\tilde{\mathbf{x}} = \mathbf{0} \quad \Rightarrow \quad \ddot{\mathbf{x}} = \ddot{\mathbf{x}}_r + \mathbf{K}_d\dot{\tilde{\mathbf{x}}} + \mathbf{K}_p\tilde{\mathbf{x}}. \tag{7.38}$$

From equation (7.34) we express $\ddot{\mathbf{q}}$ taking into account the equality $\mathbf{y} = \ddot{\mathbf{q}}$

$$\mathbf{y} = \mathbf{J}^{-1}(\mathbf{q})\left(\ddot{\mathbf{x}} - \dot{\mathbf{J}}(\mathbf{q},\dot{\mathbf{q}})\dot{\mathbf{q}}\right). \tag{7.39}$$

By replacing $\ddot{\mathbf{x}}$ in equation (7.39) with expression (7.38), the control algorithm based on inverse dynamics in the external coordinates is obtained

$$\mathbf{y} = \mathbf{J}^{-1}(\mathbf{q})\left(\ddot{\mathbf{x}}_r + \mathbf{K}_d\dot{\tilde{\mathbf{x}}} + \mathbf{K}_p\tilde{\mathbf{x}} - \dot{\mathbf{J}}(\mathbf{q},\dot{\mathbf{q}})\dot{\mathbf{q}}\right). \tag{7.40}$$

The control scheme encompassing the linearization of the system based on inverse dynamics (7.31) and the closed loop control (7.40) is shown in Figure 7.13.

7.3 Control of the contact force

The control of position is sufficient when a robot manipulator follows a trajectory in free space. When contact occurs between the robot end-effector and the environment, position control is not an appropriate approach. Let us imagine a robot

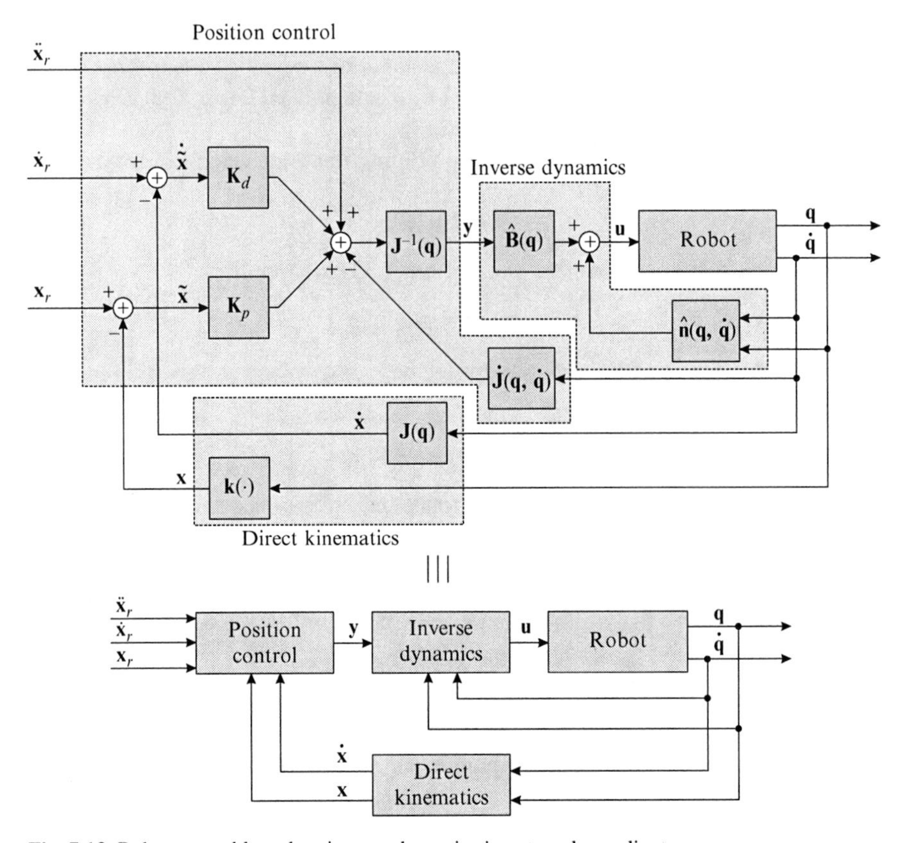

Fig. 7.13 Robot control based on inverse dynamics in external coordinates

manipulator cleaning a window with a sponge. As the sponge is very compliant, it is possible to control the force between the robot and window by controlling the position between the robot gripper and the window. If the sponge is sufficiently compliant and when we know the position of the window accurately enough, the robot will appropriately accomplish the task.

If the compliance of the robot tool or its environment is smaller, then it is not so simple to execute the tasks which require contact between the robot and its environment. Let us now imagine a robot scraping paint from a glassy surface while using a stiff tool. Any uncertainty in the position of the glassy surface or malfunction of the robot control system will prevent satisfactory execution of the task; either the glass will break, or the robot will uselessly wave in the air.

In both robot tasks, i.e. cleaning a window or scraping a smooth surface, it is more reasonable that instead of position of the glassy surface we determine the force that the robot should exert on the environment. Most of the modern industrial robots are carrying out relatively simple tasks, such as spot welding, spray painting and various point-to-point operations. Several robot applications, however, require control of the contact force. A characteristic example is grinding or a similar robot machining task. An important area of industrial robotics is also robot assembly, where several component parts are to be assembled. In such robot tasks, sensing and controlling the forces is of utmost importance.

Accurate operation of a robot manipulator in an uncertain, non-structured and changeable environment is required for efficient use of robots in an assembly task. Here, several component parts must be brought together with high accuracy. Measurement and control of the contact forces enable the required positional accuracy of the robot manipulator to be reached. As relative measurements are used in robot force control, the absolute errors in positioning of either the manipulator or the object are not as critical as in robot position control. When dealing with stiff objects, already small changes in position produce large contact forces. Measurement and control of those forces can lead to significantly higher positional accuracy of robot movement.

When a robot is exerting force on the environment, we deal with two types of robot tasks. In the first case we would like the robot end-effector to be brought into a desired pose while the robot is in contact with the environment. This is the case of robot assembly. A characteristic example is that of inserting a peg into a hole. The robot movement must be of such nature that the contact force is reduced to zero or to a minimal value allowed. In the second type of robot task, we require of the robot end-effector to exert a predetermined force on the environment. This is the example of robot grinding. Here, the robot movement depend on the difference between the desired and the actually measured contact force.

The robot force control method will be based on control of the robot using inverse dynamics. Because of the interaction of the robot with the environment, an additional component, representing the contact force $\mathbf{f}$, appears in the inverse dynamic model. As the forces acting at the robot end-effector are transformed into the joint torques by the use of the transposed Jacobian matrix (4.17), we can write the robot dynamic model in the following form

$$\mathbf{B}(\mathbf{q})\ddot{\mathbf{q}} + \mathbf{C}(\mathbf{q}, \dot{\mathbf{q}})\dot{\mathbf{q}} + \mathbf{F}_v\dot{\mathbf{q}} + \mathbf{g}(\mathbf{q}) = \tau - \mathbf{J}^T(\mathbf{q})\mathbf{f}. \tag{7.41}$$

On the right hand side of the equation (7.5) we added the component $-\mathbf{J}^T(\mathbf{q})\mathbf{f}$ representing the force of interaction with the environment. It can be seen that the force $\mathbf{f}$ acts through the transposed Jacobian matrix in a similar way as the joint torques, i.e. it tries to produce robot motion. The model (7.41) can be rewritten in a shorter form by introducing

$$\mathbf{n}(\mathbf{q}, \dot{\mathbf{q}}) = \mathbf{C}(\mathbf{q}, \dot{\mathbf{q}})\dot{\mathbf{q}} + \mathbf{F}\dot{\mathbf{q}} + \mathbf{g}(\mathbf{q}), \tag{7.42}$$

which gives us the following dynamic model of a robot in contact with its environment

$$\mathbf{B}(\mathbf{q})\ddot{\mathbf{q}} + \mathbf{n}(\mathbf{q}, \dot{\mathbf{q}}) = \tau - \mathbf{J}^T(\mathbf{q})\mathbf{f}. \tag{7.43}$$

7.3.1 Linearization of a robot system through inverse dynamics

Let us denote the control output, representing the desired actuation torques in the robot joints, by the vector **u**. Equation (7.43) can be written as follows

$$\mathbf{B}(\mathbf{q})\ddot{\mathbf{q}} + \mathbf{n}(\mathbf{q},\dot{\mathbf{q}}) + \mathbf{J}^T(\mathbf{q})\mathbf{f} = \mathbf{u}. \tag{7.44}$$

From equation (7.44) we express the direct dynamic model

$$\ddot{\mathbf{q}} = \mathbf{B}^{-1}(\mathbf{q})\left(\mathbf{u} - \mathbf{n}(\mathbf{q},\dot{\mathbf{q}}) - \mathbf{J}^T(\mathbf{q})\mathbf{f}\right). \tag{7.45}$$

Equation (7.45) describes the response of the robot system to the control input **u**. By integrating the acceleration, while taking into account the initial velocity value, the actual velocity of the robot motion is obtained. By integrating the velocity, while taking into the account the initial position, we calculate the actual positions in the robot joints. The described model is represented by the block *Robot* in Figure 7.14.

In a similar way as when developing the control method based on inverse dynamics, we will linearize the system by including the inverse dynamic model into the closed loop

$$\mathbf{u} = \hat{\mathbf{B}}(\mathbf{q})\mathbf{y} + \hat{\mathbf{n}}(\mathbf{q},\dot{\mathbf{q}}) + \mathbf{J}^T(\mathbf{q})\mathbf{f}, \tag{7.46}$$

The use of circumflex denotes the estimated parameters of the robot system. The difference between equation (7.46) and equation (7.14), representing the control based

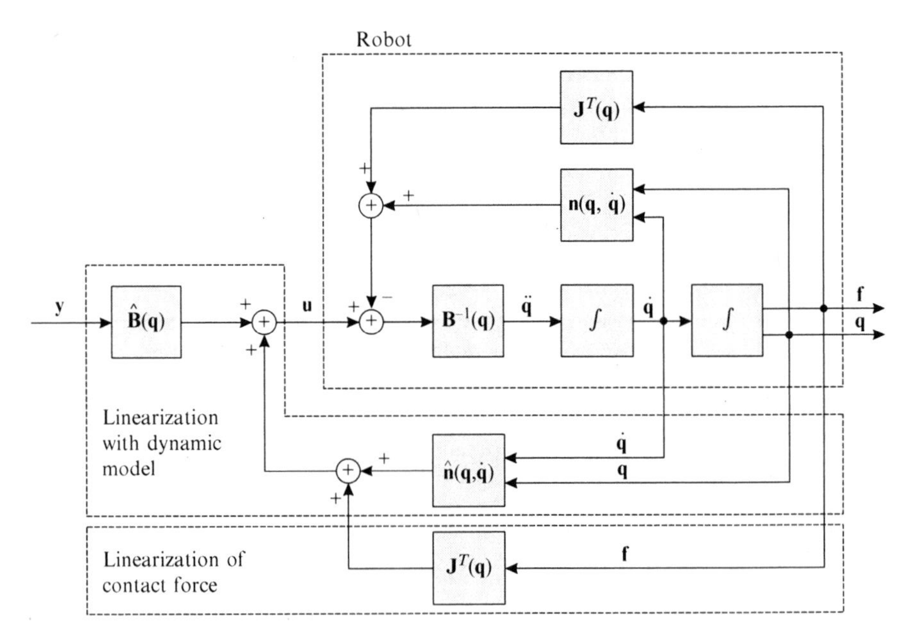

Fig. 7.14 Linearization of the control system by implementing the inverse dynamic model and the measured contact force

on inverse dynamics in internal coordinates, is the component $\mathbf{J}^T(\mathbf{q})\mathbf{f}$, compensating the influence of external forces on the robot mechanism. The control scheme, combining equations (7.45) and (7.46) is shown in Figure 7.14. Assuming that the estimated parameters are equal to the actual robot parameters, it can be observed, that by introducing the closed loop (7.46), the system is linearized because there are only two integrators between the input $\mathbf{y}$ and the output $\mathbf{q}$, as already demonstrated in Figure 7.8.

7.3.2 Force control

After linearizing the control system, the input vector $\mathbf{y}$ must be determined. The force control will be translated to control of the pose of the end-effector. This can be, in a simplified way, explained with the following reasoning: if we wish the robot to increase the force exerted on the environment, the robot end-effector must be displaced in the direction of the action of the force. Now we can use the control system which was developed to control the robot in the external coordinates (7.40). The control scheme of the robot end-effector including the linearization, while taking into account the contact force, is shown in Figure 7.15.

Up to this point we mainly summarized the knowledge of the pose control of the robot end-effector as explained in the previous chapters. In the next step we will determine the desired pose, velocity and acceleration of the robot end-effector, on the basis of the force measured between the robot end-point and its environment.

Let us assume that we wish to control a constant desired force $\mathbf{f}_r$. With the force wrist sensor, the contact force $\mathbf{f}$ is measured. The difference between the desired and measured force represents the force error

$$\tilde{\mathbf{f}} = \mathbf{f}_r - \mathbf{f}. \tag{7.47}$$

The desired robot motion will be calculated based on the assumption that the force $\tilde{\mathbf{f}}$ must displace a virtual object with inertia $\mathbf{B}_c$ and damping $\mathbf{F}_c$. In our case the virtual object is in fact the robot end-effector. For easier understanding, let us consider a system with only one degree of freedom. When a force acts on such a system, an

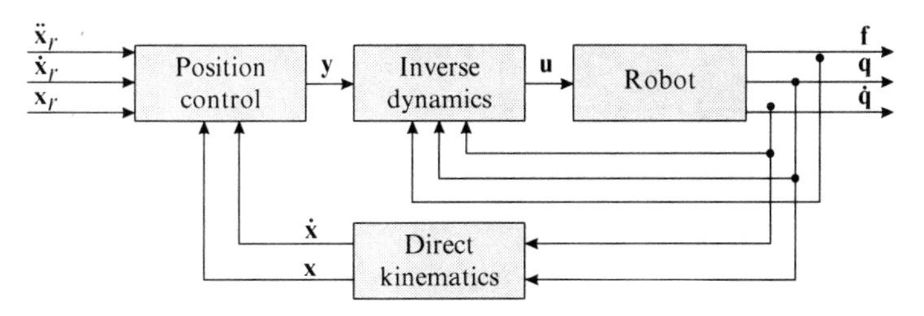

Fig. 7.15 Robot control based on inverse dynamics in external coordinates including the contact force

accelerated movement will start. The movement will be determined by the force, the mass of the object and the damping. The robot end-effector therefore behaves as a system consisting of a mass and a damper, which are under the influence of the force $\tilde{\mathbf{f}}$. For more degrees of freedom we can write the following differential equation describing the movement of the object

$$\tilde{\mathbf{f}} = \mathbf{B}_c\ddot{\mathbf{x}}_c + \mathbf{F}_c\dot{\mathbf{x}}_c. \tag{7.48}$$

The matrices $\mathbf{B}_c$ and $\mathbf{F}_c$ determine the movement of the object under the influence of the force $\tilde{\mathbf{f}}$. From equation (7.48) the acceleration of the virtual object can be calculated

$$\ddot{\mathbf{x}}_c = \mathbf{B}_c^{-1}\left(\tilde{\mathbf{f}} - \mathbf{F}_c\dot{\mathbf{x}}_c\right). \tag{7.49}$$

By integrating the equation (7.49), the velocities and the pose of the object are calculated, as shown in Figure 7.16. In this way the reference pose $\mathbf{x}_c$, reference velocity $\dot{\mathbf{x}}_c$ and reference acceleration $\ddot{\mathbf{x}}_c$ are determined from the force error. The calculated variables are inputs to the control system, shown in Figure 7.15. In this way the force control was translated into the already known robot control in external coordinates.

In order to simultaneously control also the pose of the robot end-effector, parallel composition is included. Parallel composition assumes that the reference control variables are obtained by summing the references for force control ($\mathbf{x}_c$, $\dot{\mathbf{x}}_c$, $\ddot{\mathbf{x}}_c$) and references for the pose control ($\mathbf{x}_d$, $\dot{\mathbf{x}}_d$, $\ddot{\mathbf{x}}_d$). The parallel composition is defined by equations

$$\begin{aligned}\mathbf{x}_r &= \mathbf{x}_d + \mathbf{x}_c \\ \dot{\mathbf{x}}_r &= \dot{\mathbf{x}}_d + \dot{\mathbf{x}}_c \\ \ddot{\mathbf{x}}_r &= \ddot{\mathbf{x}}_d + \ddot{\mathbf{x}}_c\end{aligned} \tag{7.50}$$

The control system incorporating the contact force control, parallel composition and control of the robot based on inverse dynamics in external coordinates is shown in Figure 7.17. The force control is obtained by selecting

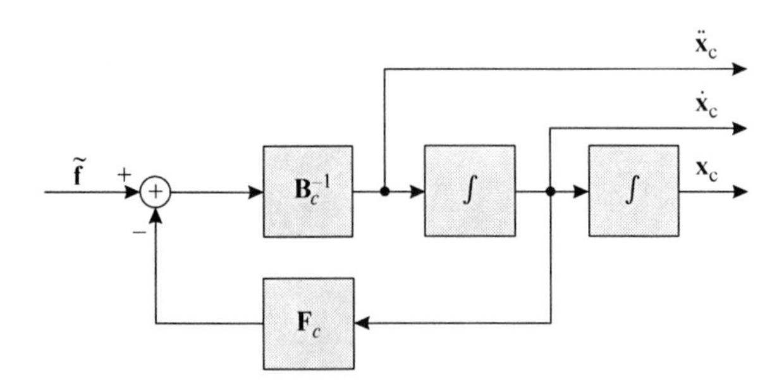

Fig. 7.16 Force control translated into control of the pose of robot end-effector

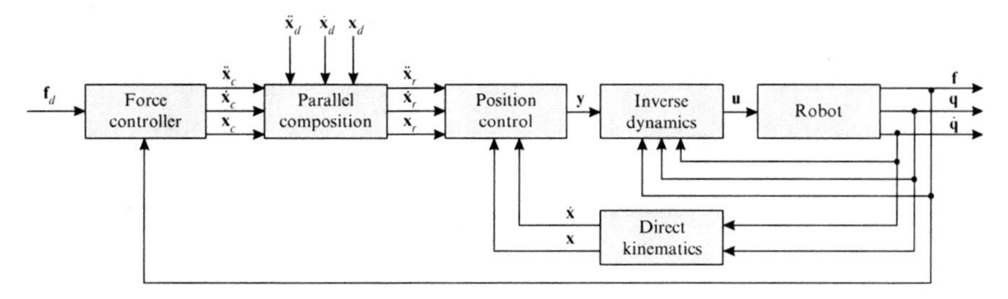

Fig. 7.17 Direct force control in the external coordinates

$$
\begin{aligned}
\mathbf{x}_r &= \mathbf{x}_c \\
\dot{\mathbf{x}}_r &= \dot{\mathbf{x}}_c \\
\ddot{\mathbf{x}}_r &= \ddot{\mathbf{x}}_c
\end{aligned}
\tag{7.51}
$$

The described control method enables the control of force. However, it does not enable independent control of the pose of the robot end-effector as it is determined by the error in the force signal.

Chapter 8
Robot environment

8.1 Robot grippers

In the same way as robot manipulators are copies of the human arm, robot grippers imitate the human hand. In most cases robot grippers are considerably simpler, as the human hand, encompassing wrist and fingers, has 22 degrees of freedom. Industrial robot grippers differ to a large extent, so it is not difficult to understand that their cost ranges from almost negligible to higher than the cost of a robot manipulator. Although a large number of various robot grippers are commercially available, it is often necessary to develop a special gripper meeting the requirements of a specific robotic task.

The most characteristic robot grippers are those with fingers. They can be divided into the grippers with two fingers and multi-fingered grippers. In industrial applications we usually encounter grippers with two fingers. The simplest two-finger grippers are only controlled between the two states, open and closed. Two-finger grippers, where the distance or force between the fingers can be controlled, are also available. Multi-fingered grippers usually have three fingers, each having three segments. Such a gripper has nine degrees of freedom which is more than robot manipulators. The cost of such grippers is high. In multi-fingered grippers the motors are often not placed into the finger joints, as the fingers can become to heavy or not strong enough. Instead, the motors are all placed into the gripper palm, while tendons connect them with the pulleys in the finger joints. Apart from grippers with fingers, we encounter in industrial robotics also vacuum, magnetic, perforation and adhesive grippers. Different end-effector tools, used in spray painting, finishing or welding, are not considered robot grippers.

Two-fingered grippers are aimed for grasping the parts in a robotic assembly process. An example of such a gripper is shown in Figure 8.1. Different end-points can be attached to the fingers in order to adapt robot grasping to the shape and the surface of the part or assembly to be grasped. With two-fingered robotic grippers pneumatic, hydraulic or electrical motors are used. Hydraulic actuation enables higher grasping forces and thus handling of heavier objects. Different structures of

T. Bajd et al., *Robotics*, Intelligent Systems, Control and Automation: Science and Engineering 43, DOI 10.1007/978-90-481-3776-3_8,

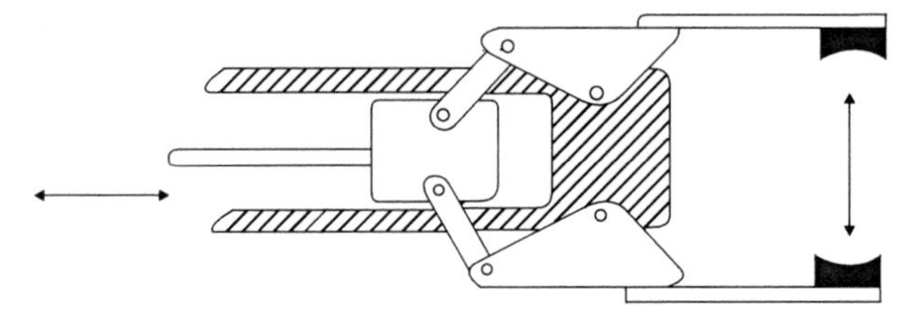

Fig. 8.1 Robot gripper with two fingers

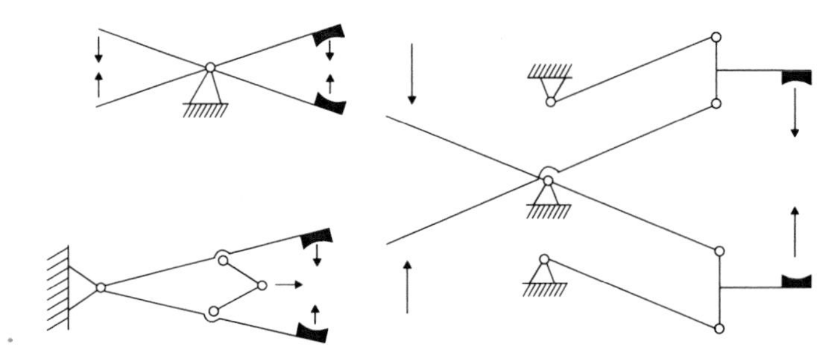

Fig. 8.2 Kinematic presentations of two-fingered grippers

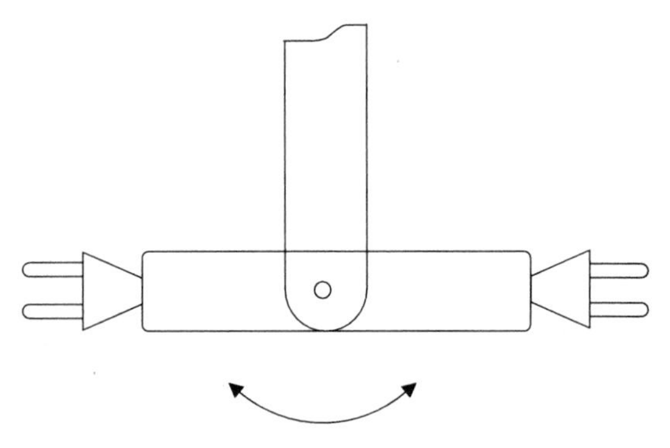

Fig. 8.3 Twofold robot gripper

two-fingered grippers are presented in Figure 8.2. Simple kinematic presentations enable the choice of an appropriate gripper for the selected task. The gripper on the right side of the Figure 8.2 enables parallel finger grasping.

In industrial processes the robot manipulators are often used for machine loading. In such cases the robot is more efficient when using a twofold gripper. The robot can simultaneously bring an unfinished part into the machine while taking a finished part out of it. A twofold gripper is shown in Figure 8.3.

Specific grippers are used for grasping of hot objects. Here, the actuators are placed far from the fingers. When handling hot objects air cooling is applied, while often the gripper is immersed into water as part of the manipulation cycle. Of utmost importance is also the choice of the appropriate material for the fingers.

When grasping lightweight and fragile objects, grippers with spring fingers can be used. In this way the maximal grasping force is constrained, while in the same time a simple way of opening and closing of the fingers is enabled. An example of a simple gripper with two spring fingers is shown in Figure 8.4.

The shape of the object requires careful design of a two-fingered robot gripper. A reliable grasp can be achieved either by form or force closure of the two fingers. Also possible is the combination of the two grasp modes (Figure 8.5).

When executing a two-fingered robot grasp, the position of the fingers with respect to the object is also important. The grasping force can be applied only to the external surfaces or only to the internal surfaces of a work-piece. Possible is also an intermediate grip where the object is grasped on internal and external surfaces (Figure 8.6).

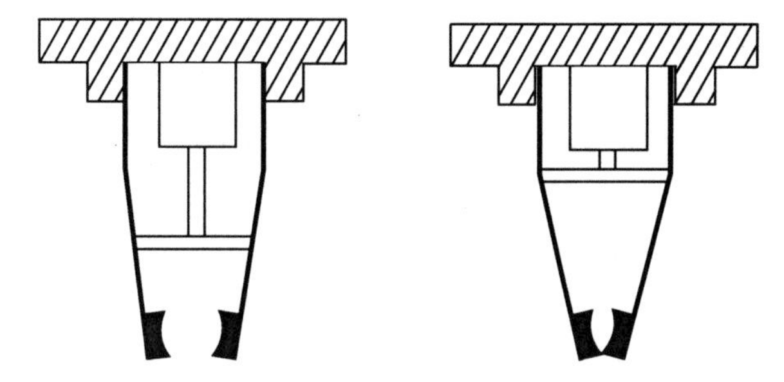

Fig. 8.4 Gripper with the spring fingers

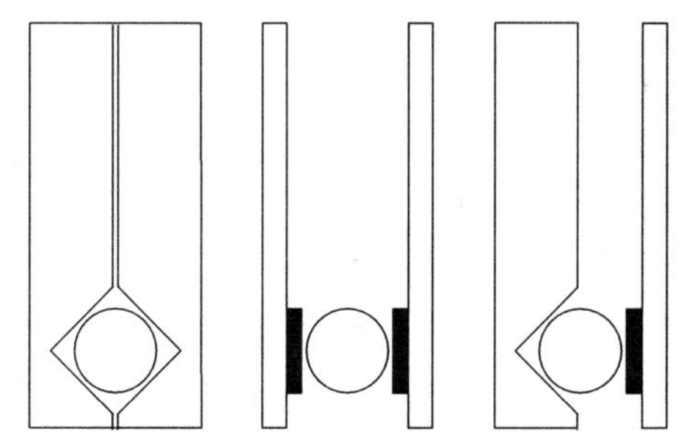

Fig. 8.5 Form closure, force closure and combined grasp

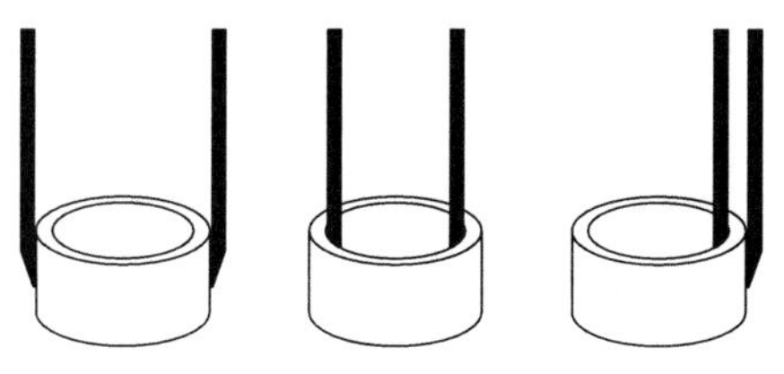

Fig. 8.6 External, internal, and intermediate grip

Among the robot grippers without fingers, the vacuum grippers are by far the most frequent. The vacuum grippers or grippers with negative pressure are successfully applied in the cases, when the surface of the grasped object is flat or evenly curved, smooth, dry and relatively clean. The advantages of these grippers are reliability, low cost and small weight. Suction heads of various shapes are commercially available. Often we use several suction heads together. We put them into a pattern that suits the shape of the object to be grasped. Figure 8.7 shows the shape of two frequently used suction heads. The head on the left is appropriate in cases when the surface is not completely smooth. Soft material of the head is adapted to the shape of the object. The small nipples of the head presented in the right side of the Figure 8.7 prevent damage of the object surface. Vacuum is produced either with Venturi or with vacuum pumps. The Venturi pump needs more power and produces only 70% vacuum. However, it is often used in industrial processes because of its simplicity and low cost. Vacuum pumps provide 90% vacuum and produce considerably less noise. In all grippers, fast grasping and releasing of the objects is required. Releasing of the very lightweight and sticky objects can be critical with vacuum grippers. In this case we release the objects with the help of positive pressure as demonstrated in Figure 8.8.

Magnetic grippers are another example of grippers without fingers. With magnetic robot grippers either permanent magnets or electromagnets are used. The electromagnets are used to a larger extent. With permanent magnets releasing of the object represents a difficulty. The problem is solved by the use of a specially planned trajectory of the end-effector where the object is retained by a fence in the robot workspace. Also in magnetic grippers several magnets are used together, placed into

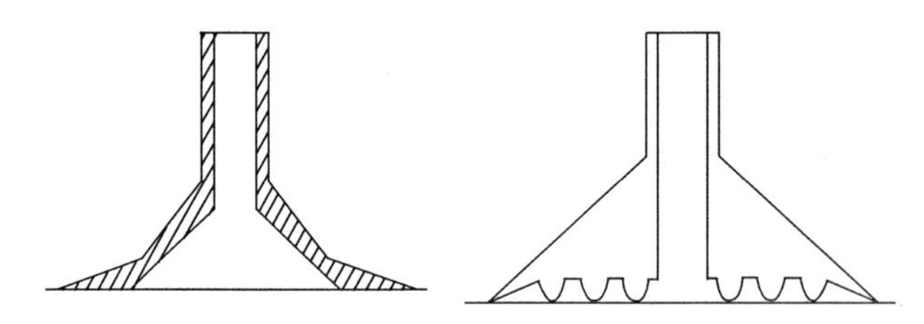

Fig. 8.7 Suction heads of vacuum gripper

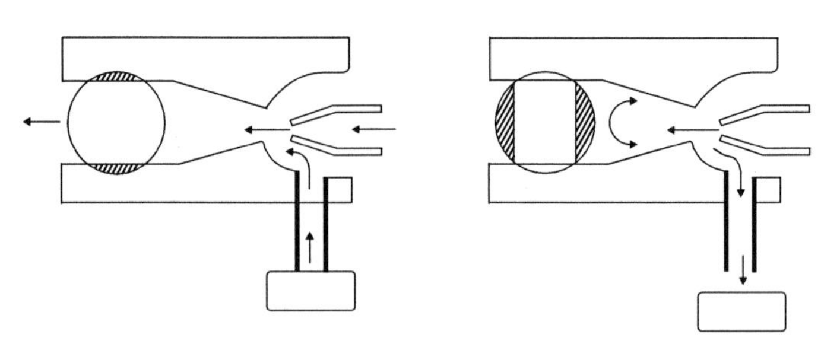

Fig. 8.8 Grasping and releasing of an object with the help of negative and positive pressure

various patterns corresponding to the shape of the object. Already small air fissures between the magnet and the object considerably decrease the magnetic force. The surfaces of the objects being grasped must be therefore even and clean. Permanent magnets are used with temperatures up to 500°C, while the electromagnets only up to 100°C.

Perforation grippers are considered as special robot grippers. Here the objects are simply pierced by the gripper. Usually we are handling material such as textile or foam rubber. Such grippers can be used only in cases when perforation does not cause damage to the object. Sheets of textile can be grasped by large brushes made of stiff nylon hairs or simply of Velcro straps.

Adhesive grippers can be used when grasping very lightweight parts. Releasing of the parts must be solved by special robot end-point trajectories where the part collides with the fence in the robot workspace and is thus removed from the adhesive gripper. Sufficient adhesive force is provided by the use of adhesive tape which must move during the operation.

8.2 Feeding devices

The task of the feeding devices is to bring parts or assemblies to the robot in such a way that the robot knows their pose. Reliable operation of the feeding devices is of utmost importance in the robot cells without robot vision. The position of a part must be accurate, as the robot end-effector always moves along the same trajectory and the part is expected to be always in the same place.

The requirements for the robot feeding devices are more strict than in manual assembly. The robot feeding devices must not damage the parts, operate reliably, position the parts accurately, work at sufficient speed, require minimal time of loading and contain sufficient number of parts.

The feeding device should not cause any damages to the parts handled. The damaged parts are afterwards inserted by the robot into assemblies which cannot function properly. The cost of such damaged assemblies is higher than the cost of a more reliable feeding device. The feeding device must handle reliably all the parts whose dimensions are within tolerance limits. The feeding device must be fast enough in order to meet the requirements of the industrial robot. The feeding device should never slow down the operation of the robot cell. Also, the feeding device should require as little time as possible for loading of the parts. It is more desirable to fill a large amount of parts into the feeding device at once than inserting them manually piece by piece. The feeding devices should contain as large number of parts as possible. In this way the number of loadings required per day is reduced.

The simplest feeding devices are pallets and fixtures. From every day life we know carton or plastic pallets for eggs. The pallets store the parts, while determining their position and sometimes also orientation. In an ideal situation the same pallet is used for shipping the parts from the vendor and for later use in the consumer's robot cell. The pallets are either loaded automatically by a machine or manually. Fragile

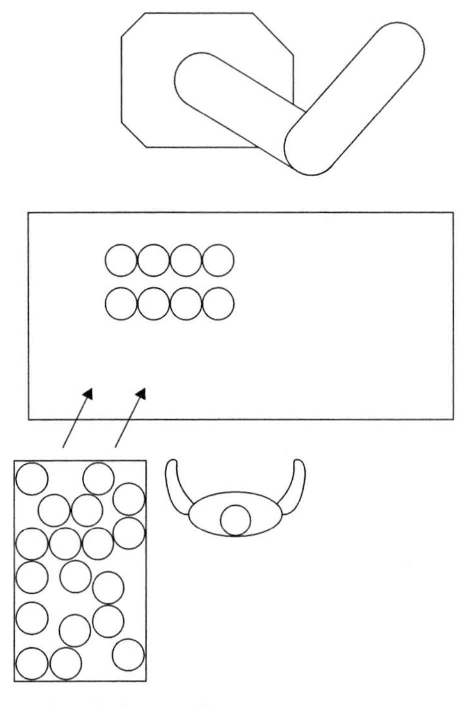

Fig. 8.9 Simultaneous loading of a fixture table

parts, flexible objects or parts with odd shape must be loaded manually. Loading of the pallets represents the weakest point of palletizing. Another disadvantage of the pallets is their rather large surface, taking up considerable area in the robot workspace.

The simplest way to bring parts into the robot cell is represented by the fixture table. The human operator takes a part from a container, where the parts are unsorted, and places it onto the fixture table inside the robot workspace (Figure 8.9). The fixture table must contain special grooves which assure reliable positioning of a part into the robot workspace. Such a fixture table is often used in welding where the component parts must be also clamped onto the table before the robot welding takes place. The time required for robot welding is considerably longer than loading and unloading what justifies the use of the fixture table.

The pallets can be loaded in advance in some other place and afterwards brought into the robot cell (Figure 8.10). In this way waiting of a robot for the human operator, who is loading the pallets, is avoided. The human worker must only bring the pallet into the robot workspace and position it properly by the use of special pins in the working table. It is important that the pallet contains a sufficient number of the parts in order to allow continuous robot operation. Exchanging the pallets in the robot workspace represents a safety problem as the operator must switch off the robot.

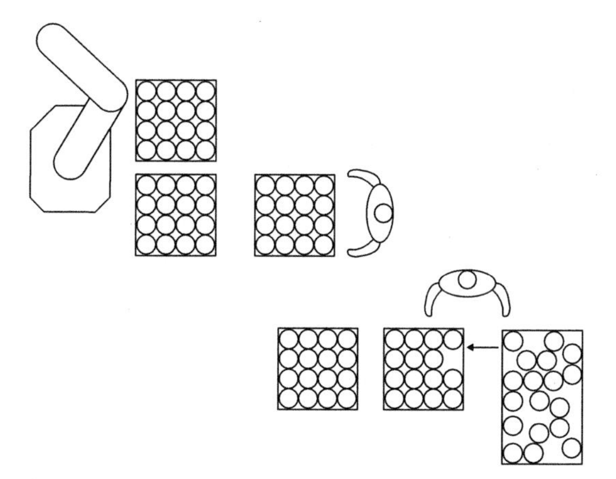

Fig. 8.10 Loading of the pallets in advance

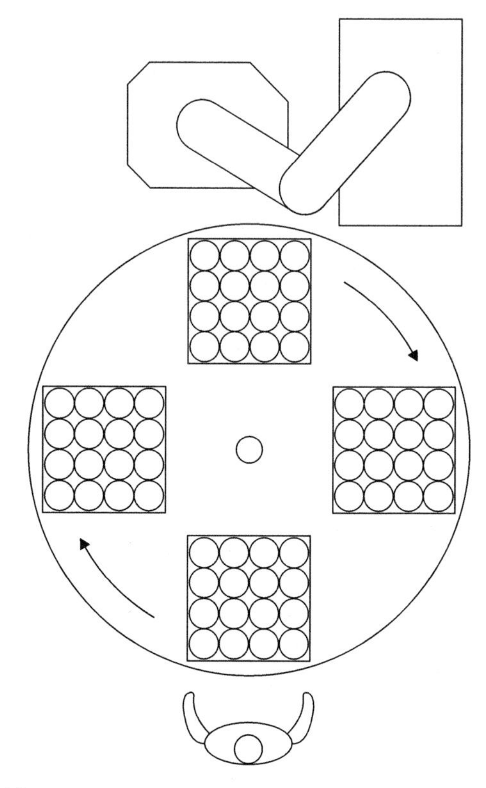

Fig. 8.11 Rotating table

A larger number of pallets can be placed on a rotating table (Figure 8.11). The rotating table enables loading of the pallets on one side, while the robot activities take place on the other side of the turntable. In this way the time of the robot cell inactivity is considerably reduced. The human operator must be appropriately protected against the movements of the robot and the rotating table. The turntables are usually actuated by electrical motors and can be positioned into a limited number of angles. The speed and repeatability of positioning are high. Closed loop controlled turntables, providing arbitrary number of positions, are also available.

We distinguish among three types of pallets: vacuum formed, injection molded and metal pallets. Since the cost of vacuum pallets is low, they are used both for packaging and shipping of the parts and for use in the robot cell. Reference holes must be made into the pallets, in order to enable (together with the pins in the worktable) simple and fast positioning of the pallets. As these pallets are the most inexpensive, it is not difficult to understand that they are the least accurate, reliable, and durable. They are made of a thin sheet of plastic material. This material is heated and by the use of vacuum formed over a mold. The inaccuracy of the pallet is the consequence of its low rigidity. Injection molded plastic pallets are used when more accurate and more durable pallets are required. The production of the mold is rather expensive, while the cost of production of a single pallet is not high. Addition of glass fibers to plastics can increase the rigidity of the pallets. We must have in mind that most vacuum and molded plastic pallets are flammable. Metal pallets are the only ones which are not flammable. They are produced by various machining approaches. The metal pallets are the most reliable and durable, while their cost is higher than the cost of plastic pallets. They are therefore only used inside the robot assembly process. When a larger number of metal pallets is necessary, they can be fabricated by casting.

Part feeders represent another interesting family of feeding devices which are not only storing the parts, but also positioning and even orienting them into the pose appropriate for robot grasping. The most common are vibratory bowl feeders (Figure 8.12). Here, the parts are disorderly filled into the bowl. The vibration of

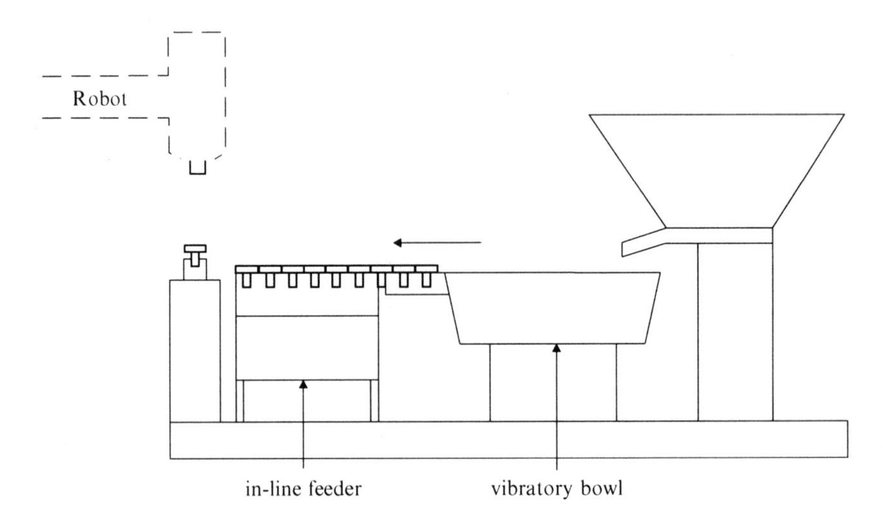

Fig. 8.12 Vibratory bowl feeder

the bowl and the in-line feeder is produced by the use of an electromagnet. Proper vibration is obtained by attaching the vibratory feeders to a large mass presented usually by a thick steel table. The vibrations cause the parts to travel out of the bowel. Specially formed spirally shaped fences force them into the required orientation. The same bowl feeder can be used for different parts, however not in the same time. Another benefit is also that the bowl holds a large number of parts while occupying only a small area in the robot workspace. The bowl feeders are not appropriate for the parts such as soft rubber objects or springs. Another disadvantage is possible damage of the parts being jammed in the bowl and handled by vibration. Also disturbing is the noise which accompanies the vibratory feeders.

A simple magazine feeder consists of a tube storing the parts and the sliding plate, which takes the parts one by one out of the magazine (Figure 8.13). The magazine is loaded manually, so that the orientation of the parts is known. Gravity pushes the parts into the sliding plate. The mechanism of the sliding plate must be designed in such a way that it prevents jamming of the parts, while only a single part is fed out from the feeder at a time. The sliding plate must block all the parts except the bottom one.

The magazine feeders are excellent solutions for handling integrated circuits (Figure 8.14). The integrated circuits are already shipped in tubes which can be used

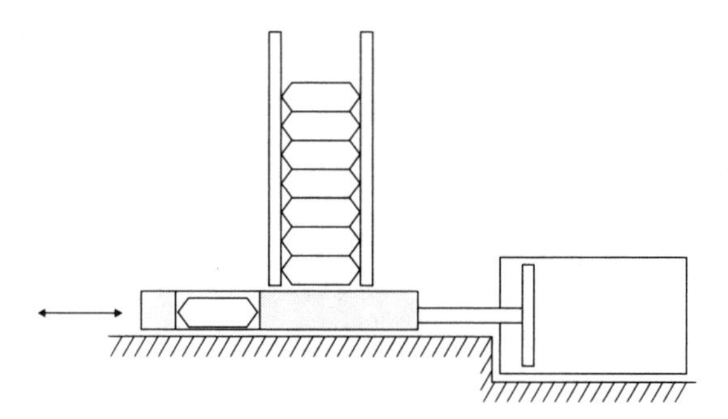

Fig. 8.13 Magazine feeder

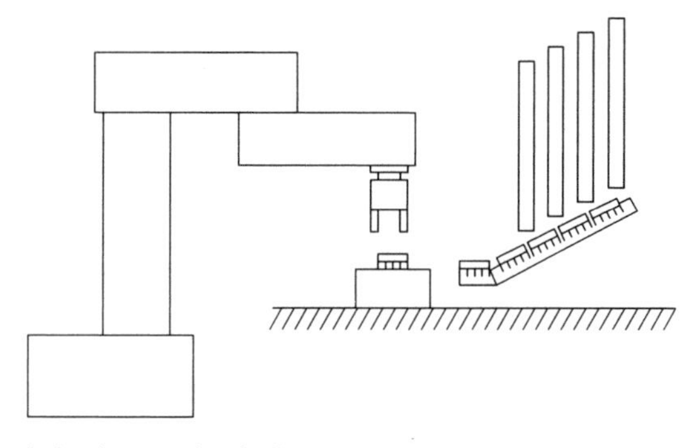

Fig. 8.14 Integrated circuit magazine feeder

for feeding purposes. The magazine feeder for integrated circuits usually consists of several tubes. The tubes are aligned along a vibratory in-line feeder. The main disadvantage of magazine feeders is manual loading. They are also not appropriate for handling of large objects.

Conveyors are used for transport of parts, assemblies or pallets between the robot cells. The simplest conveyor makes use of a plastic or metal chain which is pushing the pallets along a metal guide (Figure 8.15). An electrical motor drives the chain with constant velocity. The driving force is represented by the friction between the chain and the pallet. The pallet is stopped by special pins actuated by pneumatic cylinders. The chain continues to slide against the bottom of the pallet. When another pallet arrives, it is stopped by the first one. In this way a queue of pallets is obtained in front of the robot cell.

The turn of a conveyor is made by bending the metal guide. The advantages of the sliding chain conveyor are low cost and simplicity in handling the pallets and performing the turns. The disadvantage is that perpendicular intersections cannot be made. Also, the turns must be made in wide arcs, what takes considerable floor space in the production facility. The sliding chain conveyor is best suited when used as a single loop feeding system.

With the belt-driven conveyor, the upper part of the belt is driving pallets or other objects or material (Figure 8.16). A turn or intersection is made with the help of a special device enabling lifting, transfer and rotation of pallets.

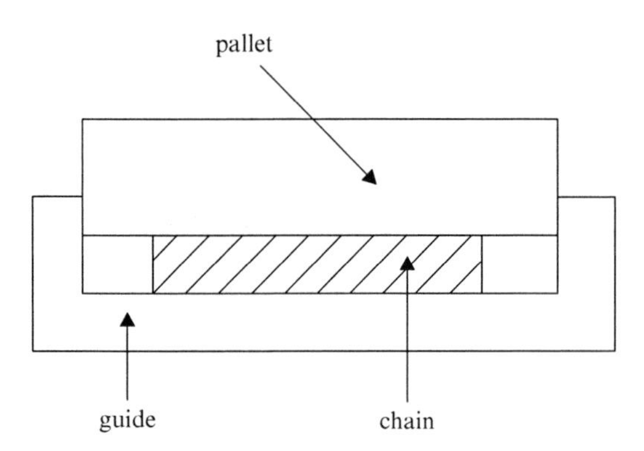

Fig. 8.15 Sliding chain conveyor (end view)

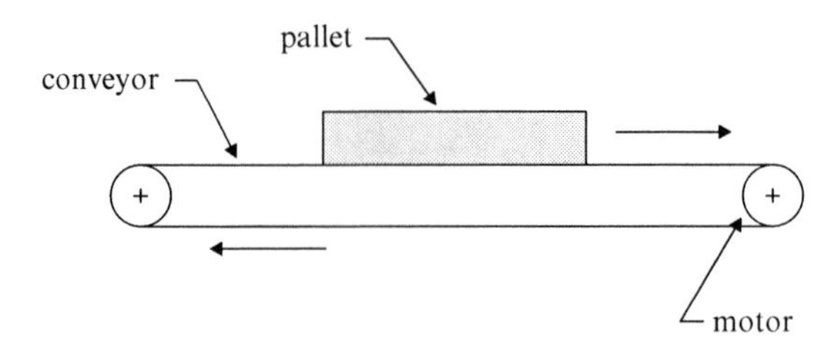

Fig. 8.16 Belt conveyor

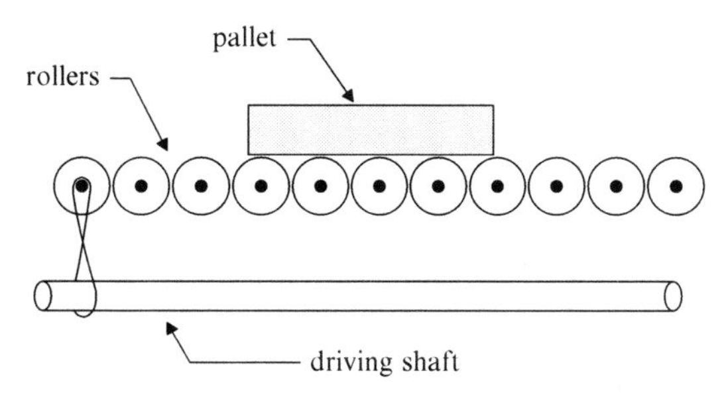

Fig. 8.17 Conveyor with rollers

The conveyor can also consist of rollers which are actuated by a common driving shaft (Figure 8.17). The driving shaft transmits torque through a drive belt to the roller shaft. The advantage of the conveyor with rollers is in low collision forces occurring between the pallets or objects handled by the conveyor. They are the consequence of low friction between the rollers and the pallets. The turns are made by the use of lift and transfer devices. The disadvantages of the conveyors with rollers are high cost and low accelerations.

8.3 Robot assembly

Robots are indispensable in today's automated industrial processes. We cannot imagine automobile industry without the long cues of welding robots. The robots are often used in cases when different objects are to be placed in different places in space, such as palettizing in e.g. food or electronic industry. Nevertheless, robots still did not fulfill all our expectations. For example, they did not replace the human operator to a sufficient extent in the processes of industrial assembly. Design of a robot assembly cell is a demanding problem requiring knowledge of robot vision, robot force control and robot grippers. These problems were dealt with in previous chapters, while here we will consider planning of the industrial assembly process. This knowledge is important when planning an intelligent control of a robot assembly cell. It can be usefully applied also when designing a product from the point of view of a simple and economical assembly.

Before starting with the formal description of a robot assembly process, several terms, definitions and assumptions must be explained. A mechanical assembly is a stable composition of interconnected parts. Each part is represented by a rigid body which does not change its shape during the assembly process. The mechanical assembly is said to be stable when all the parts maintain their relative positions during the assembly process and cannot be spontaneously disassembled. In stable mechanical assemblies all the connections between the parts are represented by plane contacts. In the plane contact without friction (e.g. a cube on an even and smooth surface) the number of the degrees of freedom, describing the relative displacements, is decreased from six to three (two translations and one rotation). A plane contact occurs also when inserting a cylindrical peg into a round hole. In this case we deal with only two degrees of freedom (translation and rotation). The number of degrees of freedom is further reduced with a prismatic peg and a rectangular hole, when only a single degree of freedom (translation) occurs. In the assembly process we encounter also attachments, which reduce the number of the degrees of freedom to zero.

A subassembly is a nonempty set of parts having either only a single element or it consists of several parts, each part having at least one plane contact with its neighbor part. The same pair of parts can be assembled in several different ways. In further consideration we will assume that the same two parts can only be joined in a unique way. In this way the functionality of a subassembly is introduced.

A sequence of a robot assembly process is represented by a succession of tasks, each task describing assembly of subassemblies into a larger subassembly. The assembly process is started when all parts are separated and ends when all parts are properly assembled into the desired mechanical assembly. In our further analysis we shall assume that in each robot task only two subassemblies are joined and remain connected throughout the assembly process.

A graph of connections $[P, C]$ represents the basic description of the assembly process. The connections are here considered as plane contacts or attachments between two neighbor parts, which must be established in each assembly task. The connections correspond to the edges of the graph

$$C = \{c_1, c_2, \ldots, c_L\},$$

while the parts belong to the nodes of the graph

$$P = \{p_1, p_2, \ldots, p_N\}.$$

Each pair of the nodes is connected by a single connection from the set C.

For further explanation of the terms and definitions associated with the assembly process, we shall make use of a simple mechanical assembly shown in Figure 8.18. It consists of only four component parts: *Cup, Stick, Receptacle* and *Bottom*. The *Cup* and the *Bottom* are supposed to be screwed on the *Receptacle*, into which the *Stick* is inserted. The same figure presents also the corresponding graph of connections.

The state of assembly is the configuration of component parts, occurring in the beginning, during and at the end of the assembly process. The state of the assembly can be described either by the connections or by the subassemblies. When describing the state with the help of connections, the following binary vector of length L is introduced

$$\mathbf{x} = [x_1, x_2, \ldots, x_L].$$

In the above vector the ith component x_i is true T, when the ith connection is established and false F, when the ith connection is not set. The initial state of the assembly process of the simple mechanical assembly shown in Figure 8.18 is described by the following five-dimensional vector

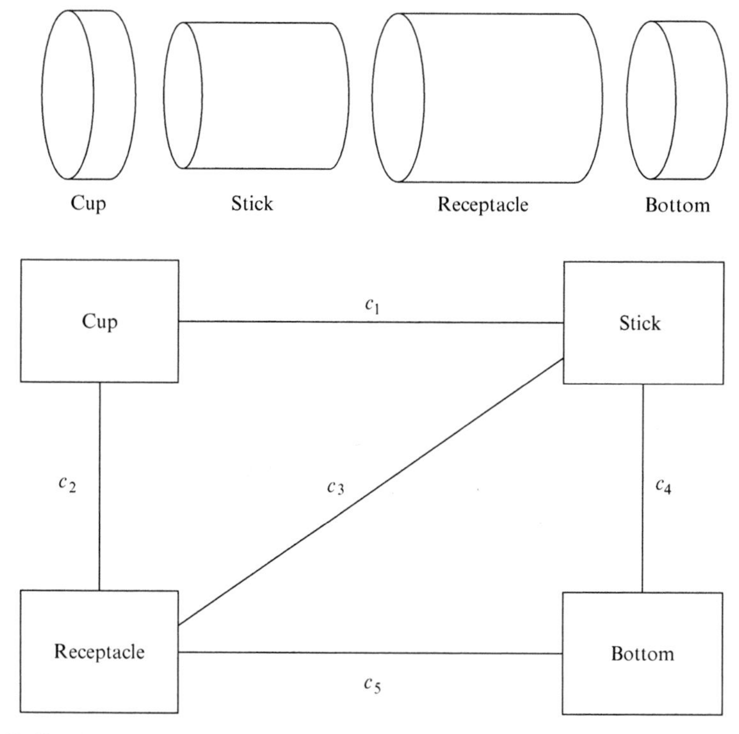

Fig. 8.18 Simple mechanical assembly and its graph of connections

$$[F, F, F, F, F].$$

The final state is

$$[T, T, T, T, T].$$

Let us suppose that the first assembly task is screwing the cup onto the receptacle. Considering the graph of connections from the Figure 8.18, this is the connection c_2. The corresponding state can be written with the following binary vector

$$[F, T, F, F, F].$$

The assembly state can be written also by the use of subassemblies. Each subassembly is described by a set of component parts. Each state is then described by the corresponding subassemblies and parts. The initial state has in our case the following form

$$\{\{Cup\}, \{Receptacle\}, \{Stick\}, \{Bottom\}\}.$$

In the beginning each component part represents a subassembly. The final state is

$$\{\{Cup, Receptacle, Stick, Bottom\}\}.$$

When the first task is to join the cup and receptacle, we can write

$$\{\{Cup, Receptacle\}, \{Stick\}, \{Bottom\}\}.$$

There exist such combinations of parts which do not represent an assembly state. Such an example is

$$\{\{Cup, Bottom\}, \{Receptacle\}, \{Stick\}\}.$$

With regard to the connection graph from the Figure 8.18 the cup and the bottom cannot be functionally joined together. In a similar way, all binary vectors do not belong to an assembly state. An example is presented by the following vector

$$[T, T, F, F, F].$$

When the connection c_1 and c_2 are established, then the connection c_3 is set automatically. The graph of connections and the descriptions of the assembly states by either binary vectors or subassemblies are the basis for more complex representations of the assembly process.

It is our aim to describe the following assembly sequence for the simple mechanical assembly shown in Figure 8.18:

- First screw the cup to the receptacle.
- Second insert the stick into the receptacle.
- Finally screw the bottom to the subassembly of the receptacle, stick and cup.

The corresponding assembly sequence can be written in one of the following four ways:

1. *Ordered list of tasks*
   The number of the elements in the list equals the number of parts minus one $(N-1)$. For our example we have three elements

$$(\{\{Cup\}, \{Receptacle\}\},$$
$$\{\{Cup, Receptacle\}, \{Stick\}\},$$
$$\{\{Cup, Receptacle, Stick\}, \{Bottom\}\}).$$

2. *Ordered list of binary vectors*
   The number of the elements in this list equals the number of parts N. In our example the list of the binary vectors has four elements

$$([F, F, F, F, F]\,[F, T, F, F, F]\,[T, T, T, F, F]\,[T, T, T, T, T]).$$

3. *Ordered list of assembly states*
   The number of the elements in this list equals the number of parts N. For our assembly sequence we write

$$(\{\{Cup\}, \{Receptacle\}, \{Stick\}, \{Bottom\}\},$$

$$\{\{Cup, Receptacle\}, \{Stick\}, \{Bottom\}\},$$

$$\{\{Cup, Receptacle, Stick\}, \{Bottom\}\},$$

$$\{\{Cup, Receptacle, Stick, Bottom\}\}).$$

4. *Ordered list of connections*
   The number of the elements in the list equals the number of parts minus one $(N-1)$. For our example we have three elements

$$(\{c_2\}, \{c_1, c_3\}, \{c_4, c_5\}).$$

These are four different ways how to formally describe a particular assembly sequence. Our further aim is to represent all possible assembly sequences for a given mechanical assembly. All assembly sequences can be described by a set of ordered lists. Many assembly sequences have common subsequences. Therefore, compact representations were proposed, encompassing all possible assembly sequences.

First, we shall consider the directed graph of assembly sequences. The directed graph represents all the assembly sequences for a mechanical assembly for which we know the graph of connections $[P, C]$. The nodes of the directed graph are presented by the stable assembly states. The edges belong to the connections or corresponding assembly tasks. The directed graph for the simple mechanical assembly in Figure 8.18 is shown in Figure 8.19. In the directed graph the initial state always appears in the following form

$$\{\{p_1\}\{p_2\} \dots \{p_N\}\},$$

while the final state equals

$$\{\{p_1, p_2, \dots, p_N\}\}.$$

Each path between the initial and the final state corresponds to a feasible assembly sequence. In such a sequence the ordered list of the edges corresponds to the ordered list of assembly tasks, while the ordered list of the nodes corresponds to the ordered list of the assembly states. With the help of the directed graph in Figure 8.19 we can see that there are ten ways, how to assemble the simple mechanical assembly of Figure 8.18. They can be described by the following ordered lists of the nodes

1. $v_1 \rightarrow v_2 \rightarrow v_7 \rightarrow v_{13}$
2. $v_1 \rightarrow v_4 \rightarrow v_7 \rightarrow v_{13}$
3. $v_1 \rightarrow v_4 \rightarrow v_{12} \rightarrow v_{13}$
4. $v_1 \rightarrow v_5 \rightarrow v_{12} \rightarrow v_{13}$
5. $v_1 \rightarrow v_3 \rightarrow v_7 \rightarrow v_{13}$
6. $v_1 \rightarrow v_3 \rightarrow v_{11} \rightarrow v_{13}$
7. $v_1 \rightarrow v_5 \rightarrow v_{11} \rightarrow v_{13}$
8. $v_1 \rightarrow v_6 \rightarrow v_{12} \rightarrow v_{13}$
9. $v_1 \rightarrow v_6 \rightarrow v_9 \rightarrow v_{13}$
10. $v_1 \rightarrow v_2 \rightarrow v_9 \rightarrow v_{13}$.

The mechanical assembly consisting of the cup, receptacle, stick and bottom is extremely simple. Let us briefly look at a more realistic example from the industrial

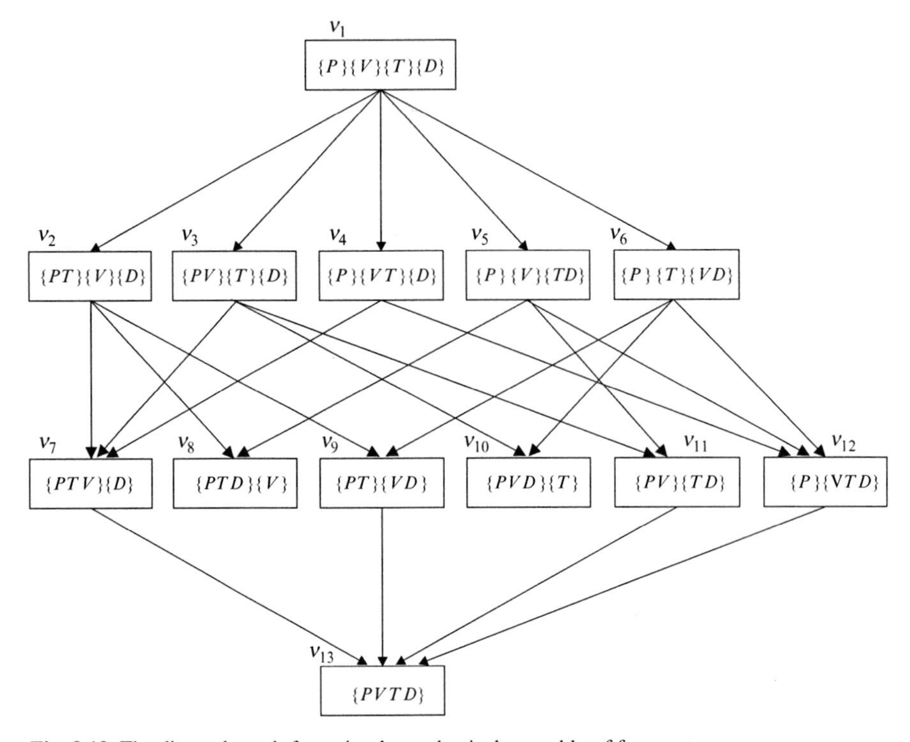

Fig. 8.19 The directed graph for a simple mechanical assembly of four parts

environment. Figure 8.20 shows a cross section of a car transmission system with 11 parts. The mechanical assembly is symmetric around the axis of revolution. The same figure presents also the corresponding graph of connections with 18 possible connections or assembly tasks. Figure 8.21 shows the directed graph which is considerably more complex than our simple assembly of four component parts. To describe the nodes of the graph the binary vectors were used instead of assembly states. Each node consists of as many little squares as there are connections. A white square represents a false F connection, while a black square belongs to a true T connection.

In further analysis of the assembly process we can assign different assembly costs to the edges (i.e. assembly tasks) of the directed graph. It is, for example, more difficult to screw the cup to the receptacle than to insert the stick into receptacle which is again more difficult than placing the stick on the bottom. Further we assume that the assemblies with a larger number of degrees of freedom are less stable and should be avoided in the robot assembly process. In our example, the subassembly of the cup and stick has three degrees of freedom, while the subassembly of the cup and receptacle has none. It is therefore wiser to first screw the cup to the receptacle and then to insert the stick, rather than to place the stick on the bottom and conclude the assembly process by screwing the receptacle to the bottom. When considering

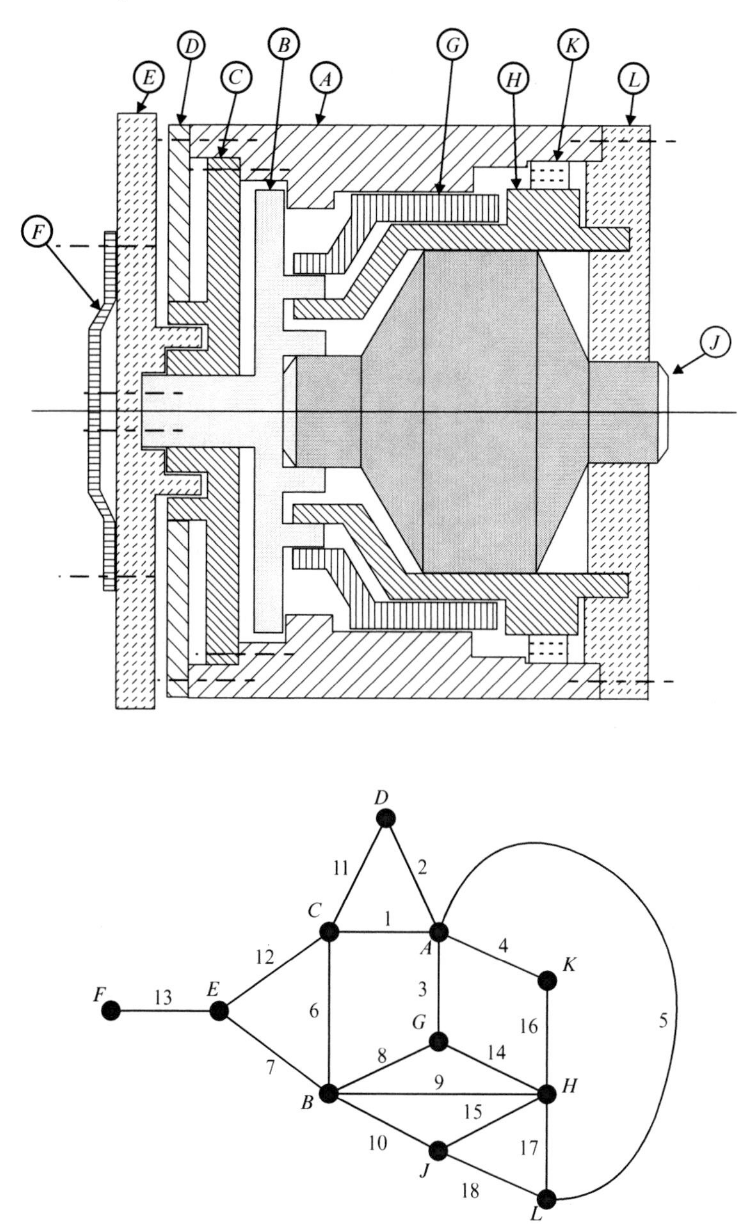

Fig. 8.20 Mechanical assembly of a car transmission system with the corresponding graph of connections

the difficulties of the assembly tasks and the stability of the assembly states, we find that some assembly sequences are more advantageous than others.

It often occurs that some tasks of the assembly sequence can be executed simultaneously. In our simple example, we can in the same time screw the cup to the

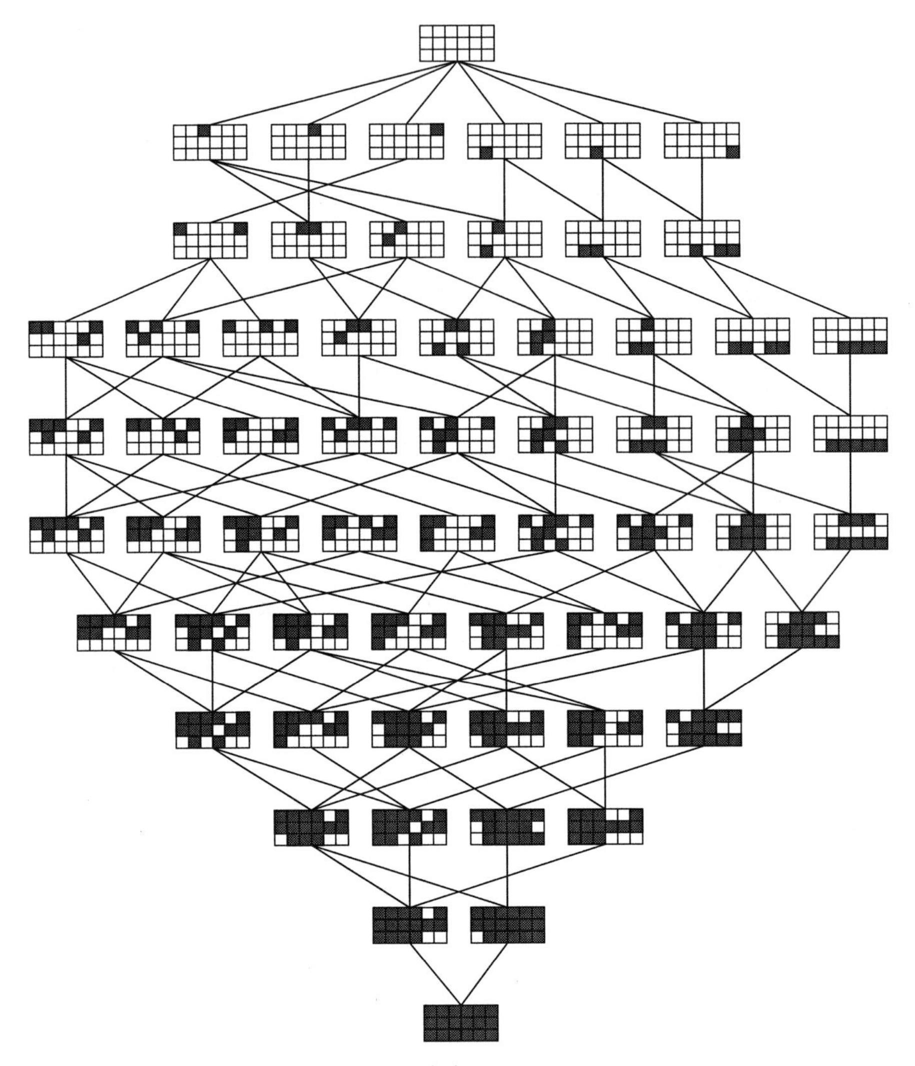

Fig. 8.21 The directed graph for a car transmission system

receptacle and from the other side insert the stick into it. In a similar way, we can first insert the stick and then simultaneously screw the cup from one side and the bottom from the other side of the receptacle. For such simultaneous assembly processes, two robot manipulators are necessary. The main advantage of the AND/OR graph, which also represents the set of all possible assembly sequences, is to show explicitly the possibility of simultaneous execution of assembly tasks. The nodes of the AND/OR graph represent the subassemblies. The arcs of the graph are feasible assembly tasks. Figure 8.22 shows the AND/OR graph belonging to the simple mechanical assembly in Figure 8.18. Every node of the graph represents a subset of parts representing a subassembly. For the sake of further explanation the nodes

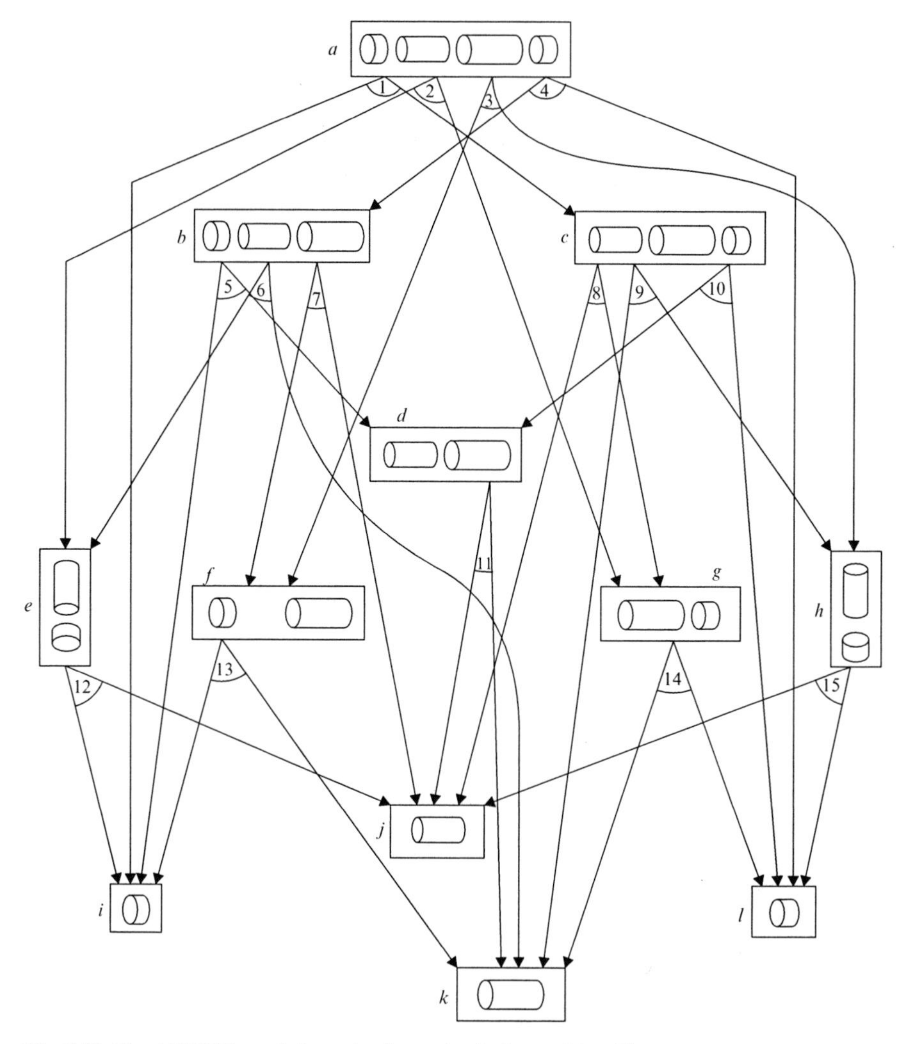

Fig. 8.22 The AND/OR graph for a simple mechanical assembly of four parts

and arcs are denoted by letters and numbers respectively. The first *a* node belongs to the set of parts describing the whole mechanical assembly. There are four arcs 1–4 leaving the first node. Each arc describes a different way to disassemble the simple mechanical assembly. The legs of each arc lead to two subassemblies. Each subassembly in the graph results from different disassembly tasks, however it appears only once in the AND/OR graph. The subassembly of the *d* node in Figure 8.22 results from two different disassembly tasks denoted by the arcs 5 and 10. Both arcs belong to different nodes. There are eight different trees, that can be drawn between the initial node *a*, belonging to the whole mechanical assembly, and the final nodes *i*, *j*, *k*, *l* representing each particular part. These trees are the solutions of

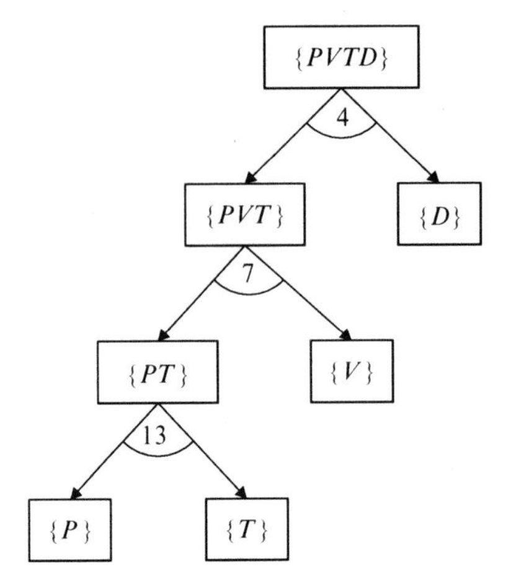

Fig. 8.23 The feasible assembly tree corresponding to the first assembly sequence

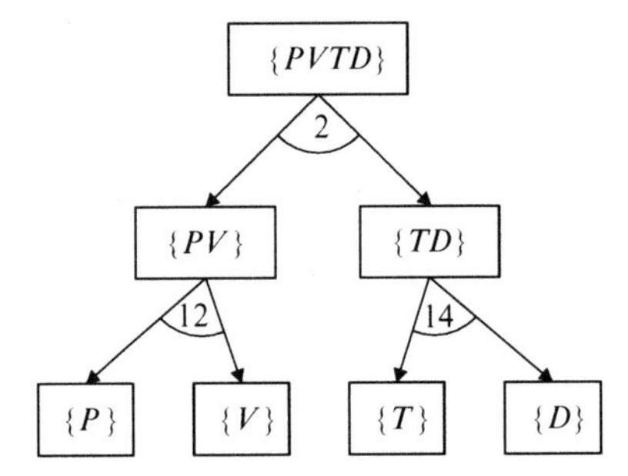

Fig. 8.24 The feasible assembly tree corresponding to the sixth and seventh assembly sequence

the AND/OR graph and represent all possible ways to assemble or disassemble the simple mechanical assembly. They are called feasible assembly trees. The assembly tree, belonging to the first assembly sequence obtained from the directed graph, is shown in Figure 8.23.

The feasible tree in Figure 8.24 belongs to the sixth and the seventh assembly sequence of the directed graph. From the graph, it is not difficult to observe that the cup and the stick can be assembled in the same time as the receptacle and the bottom. When a feasible assembly tree corresponds to more than one assembly sequence, this means that there exist such assembly sequences where particular tasks can be executed simultaneously. It is evident from the feasible tree that there is no temporal interdependence between both assembly tasks. In the above example we have shown that each assembly sequence from the directed graph belongs to a feasible assembly tree in the AND/OR graph. Each feasible assembly tree corresponds to one or more assembly sequences.

An important property of the AND/OR graph is also a significantly smaller number of nodes as compared to the directed graph. This is evident in mechanical assemblies with more than five component parts. Here we distinguish between strongly connected and weakly connected mechanical assemblies. In a strongly connected assembly, every part is connected to every other part. In a weakly connected assembly there are $N-1$ connections between the N parts. In a mechanical assembly of ten parts we deal with the following number of the nodes

strongly connected assembly	directed graph :	115975
	AND/OR graph:	1023
weakly connected assembly	directed graph :	512
	AND/OR graph:	55

In practical industrial examples we usually encounter weakly connected assemblies. The AND/OR graph, in the same way as the directed graph, presents all possible assembly sequences. The AND/OR graph shows explicitly the assembly tasks that can be executed simultaneously, which cannot be observed from the directed graph.

Chapter 9
Standards and safety in robotics

In this chapter we shall briefly consider three basic European robotic standards. The first document ISO 9946 presents the characteristics of industrial robot manipulators. The second standard ISO 9787 is entitled Coordinate Systems and Motions. The most important and most extensive is the ISO 9283 standard which describes the performance criteria and the methods for testing of industrial robot manipulators.

The first standard requires from the robot manufacturers to clearly specify the characteristics and application requests for their industrial robots. First, the main type of application should be indicated. The standard enumerates the following areas of application: material handling, assembly, spot welding, arc welding, machining, spray painting and coating, application of adhesive, tool manipulation and work inspection or verification.

In continuation the manufacturer is required to indicate the external power sources, which in robotics are electrical, hydraulic, pneumatic and combined actuators. The maximum power consumption must be also referred to. A schematic drawing of the robot mechanical structure must be presented:

- Cartesian robot (Figure 9.1)
- Cylindrical robot (Figure 9.2)
- Polar (spherical) robot (Figure 9.3)
- Anthropomorphic robot (Figure 9.4)
- SCARA robot (Figure 9.5)

In all drawings the degrees of freedom of the robot mechanism must be marked and clearly visible. The drawing must include also the base coordinate frame and the mechanical interface frame which are determined by the manufacturer and specified by the second standard.

Of special importance is the diagram showing the boundaries of the workspace (Figure 9.6). The maximal reach of the robot arm must be clearly shown in at least two planes. The range of motion for each robot axis (degree of freedom) must be indicated. The manufacturer must specify also the center of the workspace c_w, where most of the robot activities take place.

T. Bajd et al., *Robotics*, Intelligent Systems, Control and Automation: Science and Engineering 43, DOI 10.1007/978-90-481-3776-3_9,

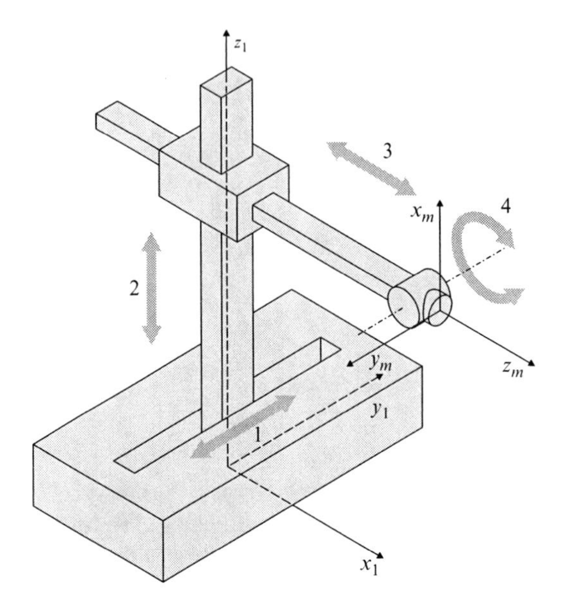

Fig. 9.1 Mechanical structure of the cartesian robot

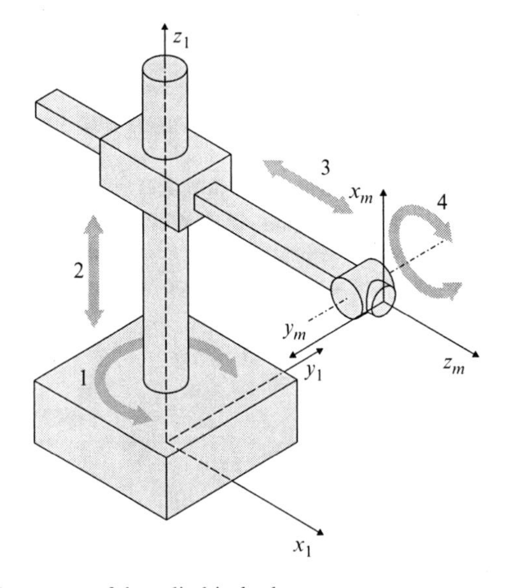

Fig. 9.2 Mechanical structure of the cylindrical robot

The robot data must be accompanied by the characteristic loading parameters, such as mass (kg), torque (Nm), moment of inertia (kgm^2) and thrust (N). The maximal velocity must be given at a constant rate, when there is no acceleration or deceleration. The maximal velocities for particular robot axes must be given with

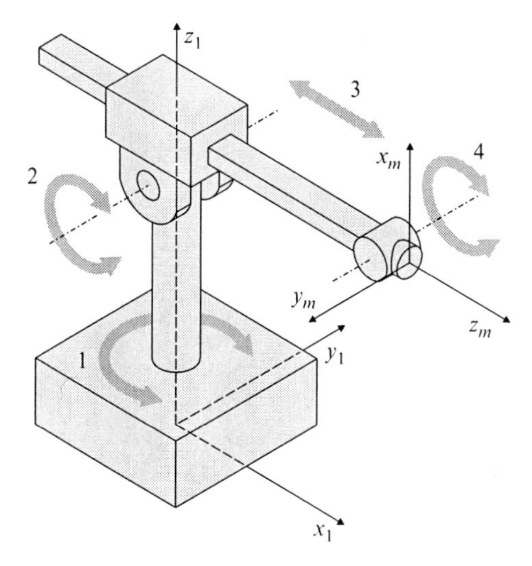

Fig. 9.3 Mechanical structure of the polar robot

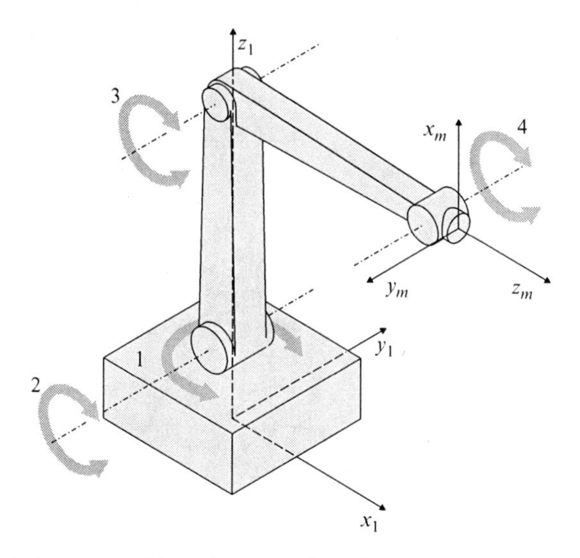

Fig. 9.4 Mechanical structure of the anthropomorphic robot

the load applied to the end-effector. The resolution of each axis movement (*mm* or °), description of the control system and the programming methods must also be presented.

The second document (ISO 9787) describes the coordinate frames and the motions of industrial robots. The standard defines three right-handed frames shown in Figure 9.7. First is the world coordinate frame x_0, y_0, z_0. The origin of the frame is defined by the user. The z_0 axis is parallel to the gravity vector, however in the

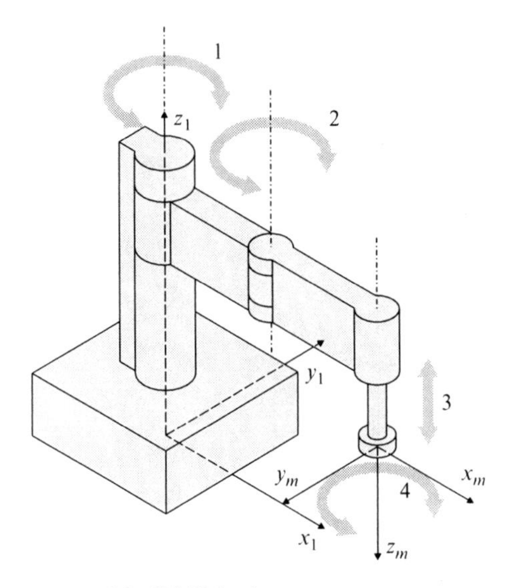

Fig. 9.5 Mechanical structure of the SCARA robot

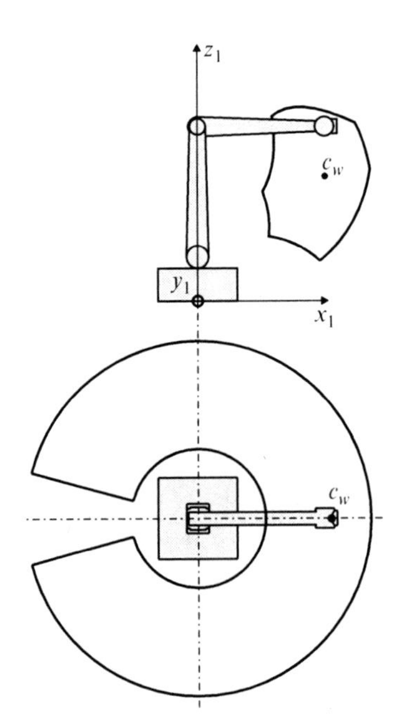

Fig. 9.6 Robot workspace

opposite direction. Second is the base coordinate frame x_1, y_1, z_1, whose origin is defined by the manufacturer. Its axes are aligned with the base segment of the robot. The positive z_1 axis is pointing perpendicularly away from the base mounting sur-

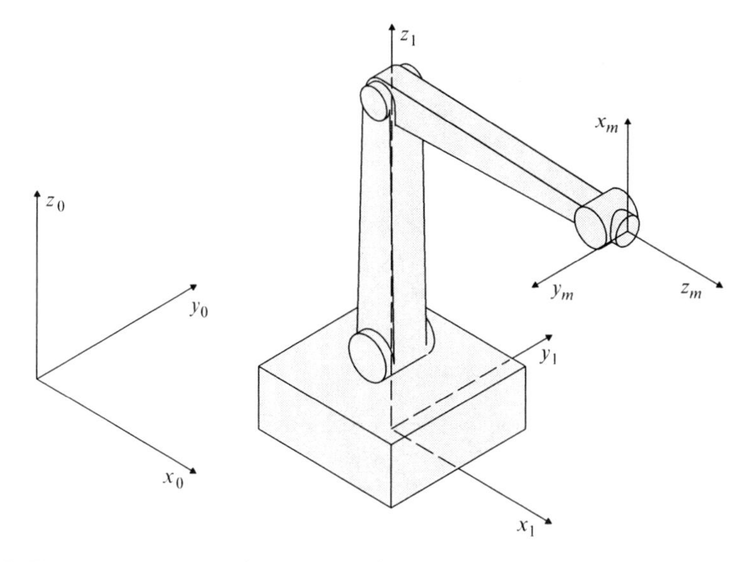

Fig. 9.7 The coordinate frames of the robot manipulator

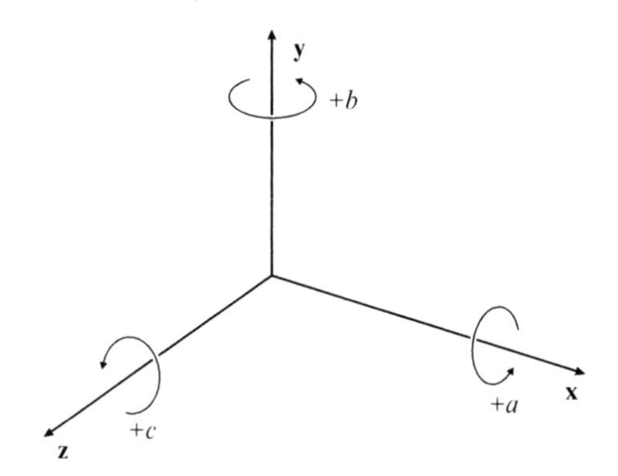

Fig. 9.8 Positive directions of translational and rotational displacements

face. The x_1 axis passes through the projection of the center of the robot workspace c_w. The frame x_m, y_m, z_m is called the mechanical interface coordinate frame. Its origin is placed in the center of the mechanical interface (robot palm) connecting the robot arm with the gripper. The positive z_m axis points away from the mechanical interface toward the end-effector. The x_m axis is located in the plane defined by the interface, which is perpendicular to the z_m axis. The standard considers also robot motions and specifies the translational and rotational displacements. The positive directions of these motions are shown in Figure 9.8.

The third standard (ISO 9283) deals with criteria and methods for testing of industrial robot manipulators. This is the most important standard as it facilitates the dialogue between manufacturers and users of the robot systems. It defines the way by which particular performance characteristics of a robot manipulator should be tested. The tests can be performed during the robot acceptance phase or in various periods of robot usage in order to check the accuracy and repeatability of the robot motions. The robot characteristics, which significantly affect the performance of a robot task, are the following:

- Pose accuracy and repeatability (pose is defined as position and orientation of a particular robot segment, usually end-effector)
- Distance accuracy and repeatability
- Pose stabilization time
- Pose overshoot
- Drift of the pose accuracy and repeatability

These performance parameters are important in the point-to-point robot tasks. Similar parameters are defined for cases when the robot end-effector moves along a continuous path. These parameters will not be considered in this textbook and can be found in the original documents.

When testing the accuracy and repeatability of a robot mechanism, two terms are important, namely the cluster and the cluster barycenter. The cluster is defined as a set of attained end-effector poses, corresponding to the same command pose. The barycenter is a point whose coordinates are the mean values of the x, y and z coordinates of all the points in the cluster. The measured position and orientation data must be expressed in a coordinate frame parallel to the base frame. The measurement point should lay as close as possible to the origin of the mechanical interface frame. Contact-less optical measuring methods are recommended. The measuring instrumentation must be adequately calibrated. The robot accuracy and repeatability tests must be performed with maximal load at the end-effector and maximal velocity between the specified points.

The standard defines the poses which should be tested. The measurements must be performed in five points, located in a plane which is placed diagonally inside a cube (Figure 9.9). Also specified is the pose of the cube in the robot workspace. It should be located in that portion of the workspace where most of the robot activities are anticipated. The cube must have maximal allowable volume in the robot workspace and its edges should be parallel to the base coordinate frame. The point P_1 is located in the intersection of the diagonals in the center of the cube. The points $P_2 - P_5$ are located at a distance from the corners of the cube equal to $10\% \pm 2\%$ of the length of the diagonal. The standard also determines the minimum number of cycles to be performed when testing each characteristic parameter:

- Pose accuracy and repeatability: 30 cycles
- Distance accuracy and repeatability: 30 cycles
- Pose stabilization time: 3 cycles
- Pose overshoot: 3 cycles
- Drift of pose accuracy and repeatability: continuous cycling during 8 h

When testing the accuracy and repeatability of the end-effector poses we must distinguish between the so called command pose and the attained pose (Figure 9.10).

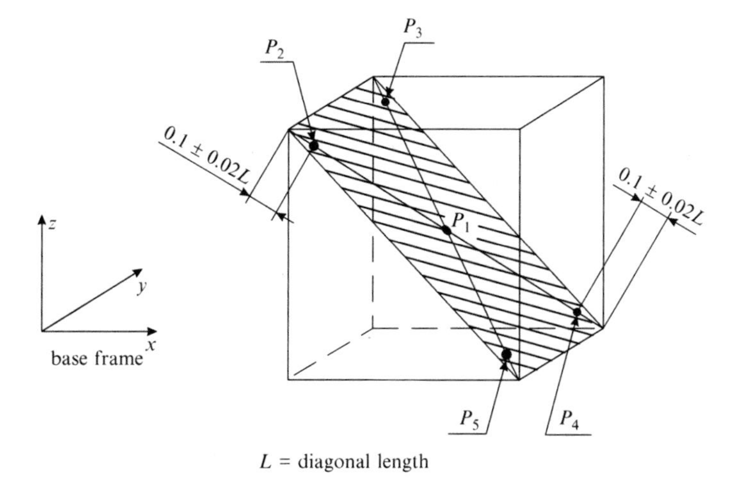

Fig. 9.9 The cube with the points to be tested

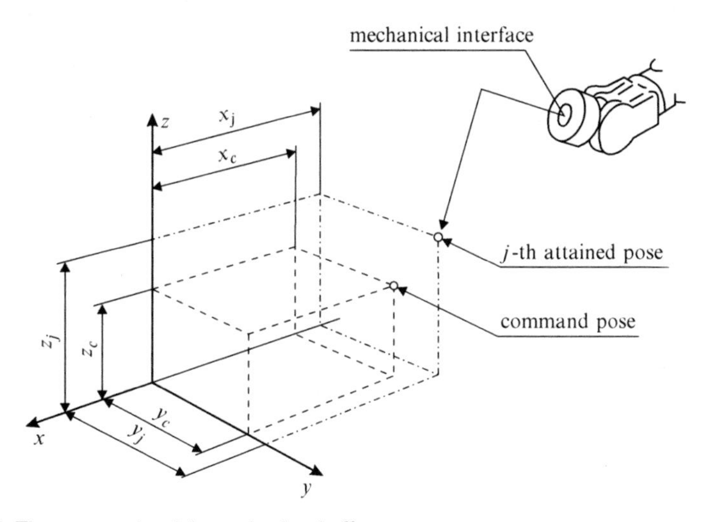

Fig. 9.10 The command and the attained end-effector pose

The command pose is the desired pose, specified through robot programming or manual input of the desired coordinates by the use of teach pendant. The attained pose is the actually achieved pose of the robot end-effector in response to the command pose. The pose accuracy evaluates the deviations, which occur between the command and the attained pose. The pose repeatability estimates the fluctuations in the attained poses for a series of repeated visits to the same command pose. The pose accuracy and repeatability are, therefore, very similar to the accuracy and repeatability of repetitive shooting at a target. The reasons for the deviations are: errors caused by the control algorithm, coordinate transformation errors, differences between the

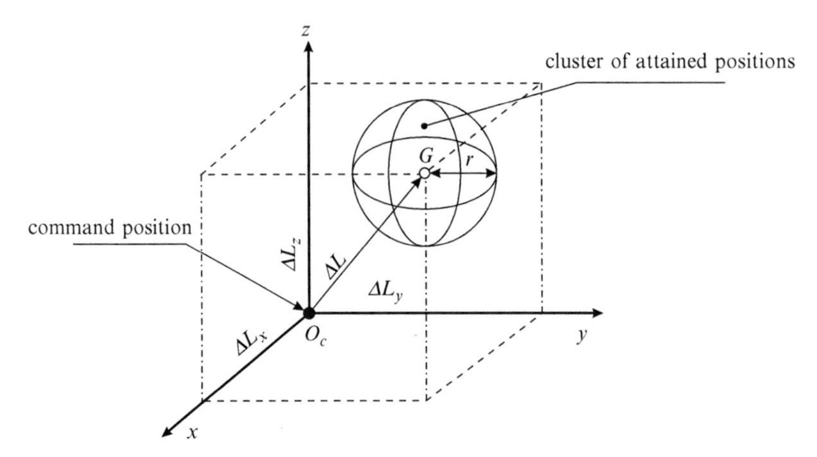

Fig. 9.11 The position accuracy and repeatability

dimensions of the robot mechanical structure and the robot control model, mechanical faults, such as hysteresis or friction, and external influences such as temperature.

The pose accuracy is defined as the deviation between the command pose and the mean value of the attained poses when the end-effector was approaching the command pose from the same direction. The position and orientation accuracy are treated separately. The position accuracy is determined by the distance between the command pose and the barycenter of the cluster of attained poses (Figure 9.11). The position accuracy is expressed by the following equation:

$$\Delta L = \sqrt{(\bar{x} - x_c)^2 + (\bar{y} - y_c)^2 + (\bar{z} - z_c)^2}. \tag{9.1}$$

In equation (9.1) $\bar{x}, \bar{y}, \bar{z}$ are the coordinates of the cluster barycenter, obtained by averaging the 30 measurement points, assessed when repeating the motions into the same command pose O_c with the coordinates x_c, y_c, z_c.

The orientation accuracy is the difference between the commanded angular orientation and the average of the attained angular orientations. It is expressed separately for each axis of the base coordinate frame. The orientation accuracy around the z axis has the following form:

$$\Delta L_c = \bar{C} - C_c. \tag{9.2}$$

In equation (9.2) $\bar{C}$ is the mean value of the orientation angles around the z axis, obtained in 30 measurements when trying to reach the same command angle C_c. Similar equations are written for the orientation accuracy around the x and y axes.

The standard exactly defines also the course of the measurements. The robot starts from point P_1 and moves into points P_5, P_4, P_3, P_2, P_1. Each point is always reached from the same direction:

$$
\begin{array}{ll}
\text{0 cycle} & P_1 \\
 & \ddots \\
\text{1st cycle} & P_5 \rightarrow P_4 \rightarrow P_3 \rightarrow P_2 \rightarrow P_1 \\
 & \ddots \\
\text{2nd cycle} & P_5 \rightarrow P_4 \rightarrow P_3 \rightarrow P_2 \rightarrow P_1 \\
 & \ddots \\
 & \ddots \\
\text{30th cycle} & P_5 \rightarrow P_4 \rightarrow P_3 \rightarrow P_2 \rightarrow P_1
\end{array}
$$

For each point the position accuracy ΔL and the orientation accuracies ΔL_a, ΔL_b and ΔL_c are calculated.

For the same series of measurements also the pose repeatability is to be determined. The pose repeatability expresses the closeness of the positions and orientations of the 30 attained poses when repeating the robot motions into the same command pose. The position repeatability (Figure 9.11) is determined by the radius of the sphere whose center is the cluster barycenter. The radius is defined as

$$
r = \bar{D} + 3S_D. \tag{9.3}
$$

The calculation of the radius r according equation (9.3) is further explained by the following equations

$$
\bar{D} = \frac{1}{n} \sum_{j=1}^{n} D_j
$$

$$
D_j = \sqrt{(x_j - \bar{x})^2 + (y_j - \bar{y})^2 + (z_j - \bar{z})^2}
$$

$$
S_D = \sqrt{\frac{\sum_{j=1}^{n} (D_j - \bar{D})^2}{n-1}}.
$$

In the above equations we again select $n = 30$, while x_j, y_j, z_j are the coordinates of the jth attained position.

The orientation repeatability for the angle around the z axis is presented in Figure 9.12. The orientation repeatability expresses how dispersed are the 30 attained angles around their average for the same command angle. It is described by the threefold standard deviation. For the angle around the z axis we have

$$
r_c = \pm 3S_c = \pm 3\sqrt{\frac{\sum_{j=1}^{n} (C_j - \bar{C})^2}{n-1}}. \tag{9.4}
$$

In equation (9.4) C_j represents the angle measured at the jth attained pose. The course of the measurements is the same as in testing of the accuracy. The radius r and the angular deviations r_a, r_b and r_c are calculated for each pose separately.

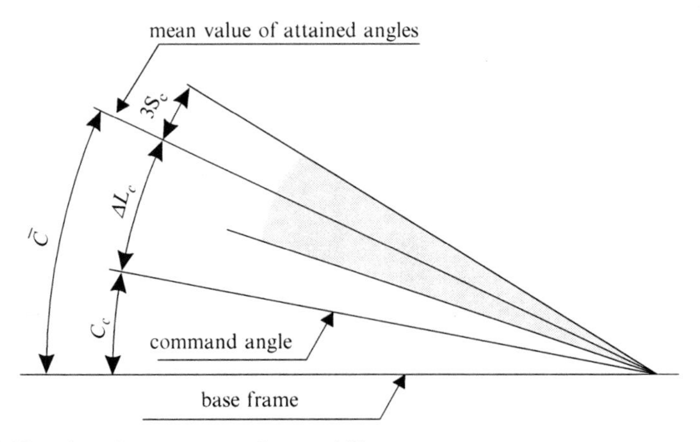

Fig. 9.12 The orientation accuracy and repeatability

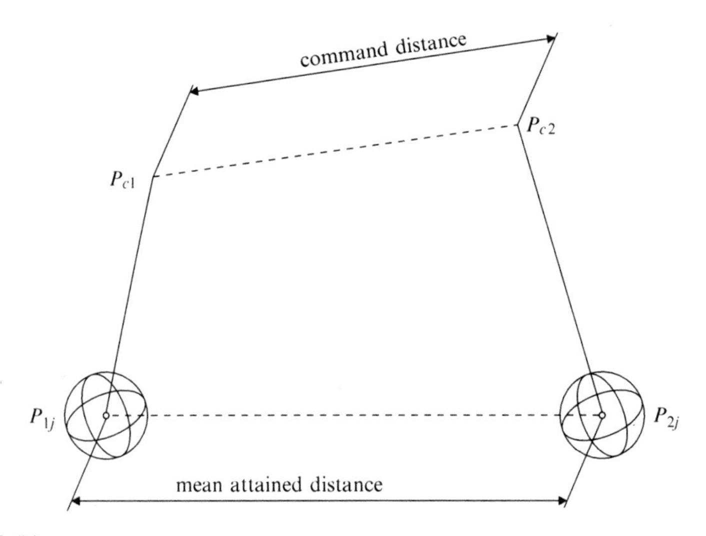

Fig. 9.13 Distance accuracy

In a similar way also the distance accuracy and repeatability are tested. The distance accuracy quantifies the deviations which occur in the distance between two command positions and two sets of the mean attained positions. The distance repeatability determines the fluctuations in distances for a series of repeated robot motions between two selected points. The distance accuracy is defined as the deviation between the command distance and the mean of the attained distances (Figure 9.13). Assuming that P_{c1} and P_{c2} are the commanded pair of positions and P_{1j} and P_{2j} are the jth pair from the 30 pairs of the attained positions, the distance accuracy is defined as

$$\Delta B = D_c - \bar{D}_D. \tag{9.5}$$

where

$$D_c = |P_{c1} - P_{c2}| = \sqrt{(x_{c1} - x_{c2})^2 + (y_{c1} - y_{c2})^2 + (z_{c1} - z_{c2})^2}$$

$$\bar{D} = \frac{1}{n} \sum_{j=1}^{n} D_j$$

$$D_j = \left|P_{1j} - P_{2j}\right| = \sqrt{(x_{1j} - x_{2j})^2 + (y_{1j} - y_{2j})^2 + (z_{1j} - z_{2j})^2}.$$

In the above equations describing the distance accuracy $P_{c1}(x_{c1}, y_{c1}, z_{c1})$ and $P_{c2}(x_{c2}, y_{c2}, z_{c2})$ represent the pair of desired positions while $P_{1j}(x_{1j}, y_{1j}, z_{1j})$ and $P_{2j}(x_{2j}, y_{2j}, z_{2j})$ are the pair of attained positions. The distance accuracy test is performed at maximal loading of the robot end-effector, which must be displaced 30 times between points P_2 and P_4 of the measuring cube. The distance repeatability is defined as

$$R_B = \pm 3\sqrt{\frac{\sum_{j=1}^{n}(D_j - \bar{D})^2}{n-1}}. \tag{9.6}$$

Let us consider another four characteristic parameters which should be tested in industrial robots moving from point to point. The first is the pose stabilization time. The stabilization time is the time interval between the instant when the robot gives the "attained pose" signal and the instant when either oscillatory or damped motion of the robot end-effector falls within a limit specified by the manufacturer. The definition of the pose stabilization time is evident from Figure 9.14. The test is performed at maximal loading and velocity. All five measuring points are visited in the following order $P_1 \rightarrow P_2 \rightarrow P_3 \rightarrow P_4 \rightarrow P_5$. For each pose the mean value of three cycles is calculated.

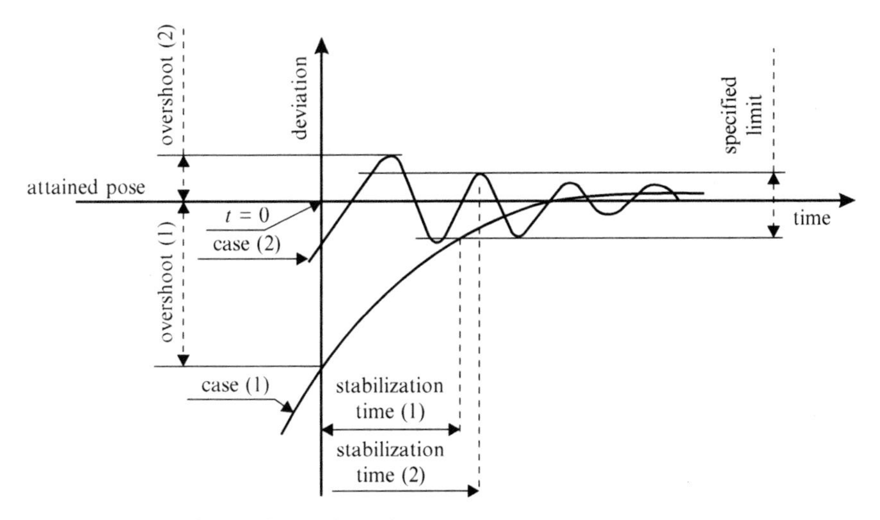

Fig. 9.14 Pose stabilization time and overshoot

A similar parameter is the pose overshoot, also shown in Figure 9.14. The pose overshoot is the maximum deviation between the approaching end-effector trajectory and the attained pose after the robot has given the "attained pose" signal. In Figure 9.14 a negative overshoot is shown in the first and positive in the second example. The instant $t = 0$ is the time when the "attained pose" signal was delivered. The measuring conditions are the same as when testing the stabilization time.

The last two parameters, to be tested in the industrial robot manipulator moving from point to point, are drift of the pose accuracy and the drift of the pose repeatability. The drift of the position accuracy is defined as

$$L_{DR} = \left| \Delta L_{t=0} - \Delta L_{t=T} \right|, \tag{9.7}$$

while the drift of the orientation accuracy is equal to

$$L_{DRC} = \left| \Delta L_{c,t=0} - \Delta L_{c,t=T} \right|. \tag{9.8}$$

The drift of the position repeatability is defined by the following equation

$$r_{DR} = r_{t=0} - r_{t=T}. \tag{9.9}$$

and the drift of the orientation repeatability is for the rotation around the z axis defined as

$$r_{DRC} = r_{c,t=0} - r_{c,t=T}. \tag{9.10}$$

The measurements are performed at maximal robot loading and velocity. The robot is cyclically displaced between points P_4 and P_2. The cyclic motions last for 8 h. Measurements are only taken in point P_4.

At the end of this chapter let us get acquainted with the safety problems in industrial robotics. As a matter of fact, robots represent an ideal solution for many industrial safety and health problems, mainly because they are capable of performing hard and fatiguing tasks in a dangerous environment. Welding and painting robots enable human workers to avoid toxic fumes and vapors. Robots also load power presses, which were frequent causes of injuries in the past. Robots work in foundries and radioactive environments. With the increasing number of robots in industrial processes, there is, however, an increased danger introduced by the robots themselves.

Industrial robots are strong devices which move quickly in their workspace. An accident in most cases occurs only when a human worker enters the robot workspace. A person steps into the robot vicinity either accidentally or even without knowing or with the aim of robot reprogramming or maintenance. It is often difficult for a human operator to judge what will be the robot's next move. Particularly dangerous are the unexpected robot motions, which are the consequence either of a robot failure or of a programming error. Many governmental organizations and large companies, together with robot producers, have developed several safety standards.

The approaches assuring safe cooperation of human workers and industrial robots can be divided into three major groups:

1. Robot safety features
2. Robot workspace safeguard
3. Personnel training and supervision

Today's robots have safety features to a large extent already built in for all three modes of operation: normal work, programming and maintenance. Some safety features are characteristic of all commercially available robots, while some are specific for a particular robot producer. Fault avoidance features increase robot reliability and safety. Such a feature, for example, prevents the robot from reaching into the press before it is open. The safety features built into the robot control unit usually enable synchronization between the robot and other machines in the robot environment. Checking the signals, indicating when a device is ready to take an active part in the robot cell, must be part of safe robot programming. The use of reliable sensors plays an important role when checking the status of machines in the robot working area. Important safety features of any robot system are also software and electric stops.

When programming or teaching a robot, the human operator must be in the robot working area. In the programming phase the velocity of the robot motions must be considerably lower than during normal work. The speed of the robot must be reduced to such a value that the human operator can avoid unexpected robot motions. The recommended maximal velocity of the robot, when there is a human worker inside the workspace, is 25 cm/s.

The teach pendant unit can be a critical component in safe robot operation. Programming errors during teaching of a robot often cause unexpected robot motions. The design of a teach pendant unit can have a significant impact on safe operation. The use of joystick control was found safer than the use of control pushbuttons. The size of emergency pushbuttons also has an important influence on the human operator's reaction times.

Special safety features facilitate safe robot maintenance. Such a feature is, for example, the possibility of switching on the control system, while the robot arm is not powered. Another feature enables passive manual motion of the robot segments, while the robot actuators are switched off. Some robot features cause the robot to stop as soon as possible, while some allow the control system to execute the current command and stop afterwards.

Most robot accidents occur when persons intentionally or carelessly enter the robot working area. The robot workspace safeguards prevent such entrance into the robot cell. There are three major approaches to the robot workspace protection:

- Barriers and fences
- Presence sensing
- Warning signs, signals and lights

Most commonly metal barriers or fences are used in order to prevent unauthorized workers from entering the robot working area. The color of the fence plays an important role, efficiently warning the not informed personnel. The fences are also an adequate protection against various vehicles that are used for transporting the material in the production hall. Safe opening of the gates, which enable entrance into the fenced up area, must also be provided. A human operator can only enter when completely switching off the robot system by the use of a control panel outside the barriers. Well designed safeguarding barriers may also protect from eventual objects which could fly out from the robot grasp.

Important safeguarding is provided by the devices detecting the presence of a person in the robot working area. These can be pressure-sensitive floor mats, light curtains, end-effector sensors, various ultrasound, capacitive, infrared or microwave sensors inside the robot cell and computer vision. The instrumented floor mats or light curtains detect the entrance of a person into the robot working area. In such a case a warning signal is triggered and normal robot working can be stopped. The end-effector sensors detect the unexpected collisions with the objects in the robot environment and cause an emergency stop. Contact-less sensors and computer vision detect various intrusions into the robot working area.

Warning signs, signals and lights can to a large extent increase the safe operation of robot cells. The warning signs alert the operators to the presence of a hazardous situation. Instruction manuals and proper training are important for efficient use of warning signs. The role of the warning signs is more effective with people who unintentionally enter the robot working area, and less effective with the operators who are familiar with the operation of the robot cell. The experienced operators often neglect the warnings and intentionally enter the robot workspace without switching off the robot with the aim to save some small amount of time. Such moves are often causes of accidents. False alarms may also reduce the effectiveness of warnings.

Selection of qualified workers, safety training and proper supervision are the prerequisites for safe working with robots. Specially critical events are startup and shutdown of a robot cell. Similarly dangerous are maintenance and programming of robots. Some robot applications (e.g. welding) include specific dangerous situations which must be well known to the workers. Those employed in the robot environment must satisfy both physical and mental requirements for their job. The selection of appropriate workers is an important first step. The second step, which is equally important, is extensive safety training. Satisfactory safety is only achieved with constant supervision of the employees. Additional training is an important component of the application of industrial robots. In the training courses the workers must be acquainted with the possible hazards and their severity. They must learn how to identify and avoid hazardous situations. Common mistakes that are causes of accidents should be explained in detail. Such training courses are usually prepared with the help of robot manufacturers.

It is expected that future robots will not work behind the safety guards with locked doors or light barriers. Instead they will be working in close cooperation with humans which leads to the fundamental concern of how to ensure a safe human–robot physical interaction. The major progress is expected in the design of lightweight flexible robot segments, compliant joints, novel actuators and advanced control algorithms.

Robot vocabulary

A

accuracy – Genauigkeit (f) – précision (f)
The ability to accurately position the robot end-effector into a preprogrammed pose. The distance between the point reached by the robot end-effector and the preprogrammed one.

actuator – Aktuator (m), Antrieb (m) – actionneur (m)
Motor converting electrical, hydraulic or pneumatic energy into power producing movement.

admittance control – Admittanzregelung (f) – commande (f) en admittance (f)
Method of control of robot manipulator which is in contact with the environment. The reference inputs to the controller are represented by the desired forces (torques) and their derivatives.

anthropomorphic robot – anthropomorpher Roboter (m) – robot (m) anthropomorphe
Robot with all joints rotational. Its movements are similar to those of the human arm.

arm, robotic – Arm (m), robotisch – bras (m) robotisé
Serial chain of segments, connected with joints powered by motors.

articulated robot – Gelenkroboter (m) – robot (m) articulé
Robot with at least two consecutive rotational joints acting around parallel axes.

assembly, robotic – Montage (f), robotische – assemblage (m) robotisé
Robot manipulation of components in order to put them together into an assembled product. Typical examples include assembly of electronic printed circuits, electric motors and computer hard drives.

automatically guided vehicle – automatisch geführter Wagen (m) – véhicule (m) à guidage automatique (m)
Robot cart without human operator. By the use of wire or rail guidance, it transports raw material, tools or finished parts over greater distances in industrial halls.

automation, robotic – Automatisierung (f), robotische – automatisation (f) robotisée
Includes kinematics, dynamics, control, simulation and programming of robot systems. Comprises also sensory systems, man-machine interfaces and elements of production technology.

autonomous robot – autonomer Roboter (m) – robot (m) autonome
Robot with ability to produce and execute its own plan and strategy of movement.

axis, rotational – Drehachse (f) – axe (m) de rotation (f)
Two robot segments enabling rotation of one segment with respect to the other.

axis, translational – Translationsachse (f) – axe (m) de translation (f)
Two robot segments enabling linear motion of one segment with respect to the other.

B

base – Fundament (n) – base (f)
The platform to which the robot arm is attached. The end of a kinematic chain opposite to the robot end-effector.

base coordinate frame – Referenzkoordinatensystem (n) – repère (m) de la base (f)
Cartesian coordinate frame attached to the robot base. Its z axis points perpendicularly out of the base.

backdrivability – Rücktreibbarkeit (f) – réversibilité (f)
Measure determining how accurately the force or movement, produced at the output of a transmission system, is transferred to its input.

C

cartesian robot – kartesischer Roboter (m) – robot (m) cartésien
Robot with three translational joints. Its workspace has the shape of prism.

collision avoidance – Kollisionschutz (m) – évitement (m) d'obstacle (m)
System of machine vision, ultrasound, infrared or microwave sensors assessing the presence of an obstacle and planning a new robot path.

coordinate measuring machine – Koordinatenmessystem (n) – machine (f) à mesurer
Passive robotic mechanism with a probe at the end-effector enabling contact or contactless assessment of distance.

compliance – Nachgiebigkeit (f) – compliance (f)
Feature of a robot allowing for small displacements due to elastic behavior between robot end-plate and the gripper or tool.

computer aided manufacturing – computergestützte Fertigung (f) – fabrication (f) assistée par ordinateur (m)
Producing a product by the use of computer technologies encompassing planning of products, tools and processes by numerically controlled machines.

contact sensor – taktiler Sensor (m), Kontaktsensor (m) – détecteur (m) de contact (m), capteur (m) tactile
Detects contact between robot end-effector and environment.

continuous path control – Bahnregelung (f) – commande (f) continue
Robot control where the robot end-point moves between closely spaced points. The continuous trajectory is achieved by interpolation.

coordinate transformation – Koordinatentransformation (f) – transformation (f) des coordonnées (f)
A 4×4 matrix used to describe pose (position and orientation) or displacement (translation and rotation) of a coordinate frame in space.

coating, robotic – Oberflächenbearbeitung (f), robotische – pulverisation (f), peinture (f) robotisée
Robot manipulation of a tool, e.g. a spray gun, to apply material such as paint to the surface of an object. Robotic coating results in more uniform application of material, reducing waste of material and reducing exposure of humans to toxic materials.

cylindrical robot – zylindrischer Roboter (m) – robot (m) cylindrique
Robot with two translational and one rotational degree of freedom. The shape of workspace is cylindrical.

D

degree of freedom – Freiheitsgrad (m) – degré (m) de liberté (f)
Number of independent coordinates (not including time) necessary for the complete description of the pose of a mechanical system. The number of independent ways the end-effector can move. Number of translational and rotational robot joints.

dexterity – Fertigkeit (f), Geschicklichkeit (f) – dextérité (f)
The ability of the robot gripper to achieve various orientations with the robot end-point in a specified position.

disassembly – Zerlegung (f) – désassemblage (f)
Process where products are decomposed into parts and subassemblies.

distal – distal – distal
Direction away from the robot base toward the robot end-effector.

dynamics, direct, inverse – Dynamik (f), direkte, inverse – dynamique (f) directe, inverse
Direct dynamics denotes calculation of robot end-point trajectories from the known joint forces and torques. Inverse dynamics is the calculation of joint forces and torques resulting in the desired robot end-point trajectories.

E

emergency stop – Nothalt (m) – arrêt (m) d'urgence (f)
Removing of the drive power from the robot actuators.

encoder – Codierer (m) – codeur (m)
Transducer converting position of a translational or a rotational joint to digital data.

end-effector – Endeffektor (m) – effecteur (m) terminal
The end of a kinematic chain opposite to the robot base. Enables attachment of a gripper or a tool such as spraying nozzle or welding gun.

end-point control – Endpunktregelung (f) – commande (m) de l'effecteur (m) terminal
Control of robot joints such that the end-point moves along a desired path.

Euler angles – Eulerwinkel (m) – angles (m) d'Euler
Three angles determining the orientation of an object in space.

exoskeleton – Exoskelett (n) – exosquelette (m)
Robot mechanism with rotational joints which can be attached to the human extremity, usually applied for teleoperation purposes.

external sensor – externer Sensor (m) – capteur (m) externe
Device which by the use of sensory information affects robot movements and is not part of the robot manipulator.

exteroception – Umgebungswahrnehmung (f) – extéroception (f)
Assessment of robot environment with external sensors.

F

finishing, robotic – Endbearbeitung (f), robotische – finition (f) robotisée
Use of an industrial robot performing continuous path movements needed for finishing tasks such as spraypainting or coating.

force closure – Kraftschluss (m) – fermeture (f) des forces (f)
The ability of the robot grasp to resist arbitrary external forces.

force-torque sensor – Kraft-Momenten sensor (m) – capteur (m) d'effort (m)
Sensor in robot wrist measuring force and torque between robot end-effector and environment in three orthogonal directions.

form closure – Formschluss (m) – fermeture (f) géométrique
Geometric property of robot grasp described by complete constraint of the grasped object.

force control – Kraftregelung (f) – commande (f) en effort (m)
Robot control with respect to the difference between the desired force and the force measured at the robot end-point.

G

gantry robot – Portalroboter (m) – robot (m) portique
Overhead mounted cartesian robot with at least three degrees of freedom. It is characterized by a large workspace and heavy payload.

grasp planning – Griffplanung (f) – planification (f) de prise (f)
Capability of a robotic system to determine where and how to grasp objects in order to provide a stable grasp.

gripper – Greifer (m) – préhenseur (m)
Gripper (usually with two fingers) grasping objects of different shape, mass and material. It is actuated by either pneumatic, hydraulic or electrical motors. It can be equipped with sensors of force or of proximity.

H

hand, robotic – Hand (f) robotische – main (f) robotisée
Robot gripper with more than three fingers, each having two or three segments. Robot hands are capable of dexterous tasks resembling those of the human hand.

hand coordinate frame – Werkzeugkoordinatensystem (n) – repère (m) de l'effecteur (m) terminal
Coordinate frame attached to the robot end-effector.

harmonic drive – Wellengetriebe (n) – réducteur (m) harmonique
System with high transmission ratio using inner and outer gear bands to provide smooth robot joint motion.

hexapod – Sechsfüßler (m) – hexapode (m)
A robot using six legs in order to walk over uneven terrains.

homogenous transformation – homogene Transformation (f) – transformation (f) homogène
Matrix 4×4 describing position and orientation of a coordinate frame with respect to the reference frame. It is used also to describe the displacement i.e. translation and rotation.

human-machine interface – Bedienungsschnittstelle (f) – interface homme (m)-machine (f)
Interface between the robot and the operator through devices such as teach pendant or computer.

humanoid – Humanoide (m) – humanoïde (m)
Robot having physical properties of a human appearance, bipedal walking, manipulation and machine vision.

hybrid control – Hybridregelung (f) – commande (f) hybride
Control of robot end-effector position with simultaneous control of the contact force between robot and environment.

hyperredundant manipulator – unterbestimmter Manipulator (m) – manipulateur (m) hyper redondant
Robot mechanism with many redundant degrees of freedom with respect to the task performed.

I

industrial robot – Industrieroboter (m) – robot (m) industriel
Industrial robot is a feedback controlled, reprogrammable, multipurpose system. It is programmable in three or more degrees of freedom.

inspection, robotic – Prüfung (f), robotische – inspection (f) robotiseé
Robot manipulation and sensory system (video camera, laser, ultrasonic detector) checking the compliance of a part or assembly with specifications.

interface, robotic – Schnittstelle (f) – interface (m) robotique
Mechanical connection between robot end-point and gripper. Mounting plate at the end of the last robot segment enabling attachment of various tools.

impedance control – Impedanzregelung (f) – commande (m) en impédance (f)
Method of control of a robot in contact with the environment. The reference inputs to the controller are the desired positions and their derivatives.

J

Jacobian matrix – Jacobimatrix (f) – matrice (f) jacobienne
Matrix of partial derivatives describing the linear relation between velocities expressed in base and joint coordinates.

joint – Gelenk (m) – articulation (f)
Contact of two surfaces which either slide (translate) or rotate.

K

kinematic singularity – kinematische Singularität (f) – singularité (f) cinématique
The kinematic singularity occurs when it is not possible to solve the inverse Jacobian matrix and thus calculate the joint velocities from the known velocities of the robot end-point. It is reflected in decreased mobility of the robot mechanism.

kinematic structure – kinematische Struktur (f) – structure (f) cinématique
Physical composition of the robot including joints, links, actuators and end-effector tools.

kinematic chain – kinematische Kette (f) – chaîne (f) cinématique
Combination of successive robot segments connected by rotational or translational joints.

kinematic pair – kinematisches Paar (n) – paire (m) cinématique
Two robot segments connected by translational or rotational degree of freedom.

kinematic model – kinematisches Modell (n) – modèle (m) cinématique
Mathematical model describing relations between trajectories, velocities and accelerations of joints and end-effector.

kinematics, direct, inverse – Kinematik (f), direkte, inverse – cinématique (f) directe, inverse
Direct kinematics calculates the robot end-effector pose (velocities, accelerations)

from the known joint positions (velocities, accelerations). Inverse kinematics calculates the joint positions (velocities, accelerations) from the known end-effector pose (velocities, accelerations).

L

laser welding, robotic – Laserschweißen (n), robotisches – soudage (m) à laser (m) robotisé
Robotic control of a light beam focused to a very small spot, where the metal melts and the weld is formed.

load capacity – Belastung (f) – capacité (f) de charge (f)
The maximal total weight that can be applied at the end of the robot arm without violating the specifications of the robot.

M

machine loading, robotic – Bestückung (f), robotische – chargement (m) robotisé
Use of robots for grasping a workpiece from e.g. conveyor belt, orienting it correctly and inserting it into a machine. After processing the robot unloads the workpiece. The greatest efficiency is usually achieved when a single robot is used to service several machines.

machining, robotic – Bearbeitung (f), robotische – usinage (m) robotisé
Robot manipulation necessary to perform drilling, grinding, routing or other similar operations.

manipulation, robotic – Manipulation (f), robotische – robotique (f) de manipulation (f)
Robotic handling of the objects by moving, inserting or orienting them, to be in the proper pose for machining or some other operation.

manipulator – Manipulator (m) – robot (m) manipulateur (m)
Mechanical aspect of the robot mechanism consisting of a series of successive segments connected by joints.

manufacturing cell – Produktionszelle (f) – cellule (f) de production (f)
Manufacturing unit consisting of robots, numerically controlled machines or workstations, transport systems and storage buffers.

material handling, robotic – Materialhandhabung (f), robotische – manutention (f) robotisée
Capability of robot to transport objects. Cooperation of robot with material handling devices, such as containers, pallets, loading bins, conveyors, guided vehicles or carousels.

mechatronics – Mechatronik (f) – mécatronique (f)
Integration of mechanical and electrical engineering with control and computer engineering with the aim to design and manufacture industrial products or processes.

medical robotics – Medizinrobotik (f) – robotique (f) médicale
Usage of robots in planning and execution of medical procedures.

micromanipulation – Mikromanipulation (f) – micromanipulation (f)
Technology of assembly of micromechanical systems.

micromechanical system – mikromechanisches System (n) – système (m) micromécanique
Mechanical components, whose size typically ranges from 10 to a few 100 μm. They are manufactured by using computer-aided design, lithographic approaches and micromachining tools. Their applications are in accelerometers, oscillators, optical components, fluidic and biomedical components.

microrobot system – Mikrorobotisches System (n) – système (m) microrobotique
Robotic system including micromanipulators, micromachines, and human-machine interfaces.

mobile robot – mobiler Roboter (m) – robot (m) mobile
Programmable wheeled robot usually moving over level surfaces.

modular robot – modularer Roboter (m) – robot (m) modulaire
Robot built of independent blocs (segments, joints, actuators), which can be combined into a variety of kinematic structures.

motion planning – Bewegungsplanung (f) – planification (f) de mouvement (m)
Planning of the path of the robot end-effector or mobile robot from initial to final point, while avoiding obstacles in the environment.

multi-robot system – Mehrrobotersystem (n) – système (m) multi-robots
Robotic system consisting of two or more robots executing a task requiring collaboration of robots.

O

orientation – Orientierung (f) – orientation (f)
Three rotational degrees of freedom of an object in space.

P

palettizing – Palettieren (n) – palettisation (f)
Loading of parts into containers keeping them in organized order.

parallel manipulator – Parallelmanipulator (m) – robot (m) parallèle
Robotic mechanism where two or more closed kinematic chains connect the end-effector to the base. Parallel manipulators are characterized with higher accuracy than serial manipulators.

path – Bahn (f) – trajectoire (f)
Trajectory of a robot end-effector or of a mobile robot when performing a specific task.

pick-and-place – Punktsteuerung (f) – prise et pose
Positioning task where the robot grasps an object at one place and releases it at another.

point-to-point control – Punkt-zu-Punktregelung (f) – commande (m) point à point (m)
Programming of robot to move from one position to the next. The intermediate path is determined by the robot controller.

pose – Stellung (f) – pose (f)
Position and orientation of a body.

position – Position (f) – position (f)
Three translational degrees of freedom describing the site of an object in space.

position control – Positionsregelung (f) – commande (m) en position (f)
Robot control where the reference signal represents the desired position of the robot end-point.

position sensor – Lagesensor (m) – capteur (m) de position (f)
Sensor detecting the position of the rotor relative to the stator of a motor.

programming of robot – Roboterprogrammierung (f) – programmation (f) de robot (m)
Development of a computer program with the instructions for robot operation.

proprioception – Propriozeption (f) – proprioception (f)
The assessment of the state of the robot system by use of internal sensors in robot joints.

proximity sensor – Näherungsensor (m) – capteur (m) de proximité (f)
Sensor detecting short distances. Proximity sensors typically work on the principle of triangulation.

proximal – proximal – proximal
Direction away from the robot end-effector toward robot base.

pushing, robotic – Schieben (n), robotisches – contrôle (m) par poussée (f)
Pushing of an object with robot fingers in order to decrease the uncertainty in the pose of the object.

R

redundant manipulator – redundanter Manipulator (m) – robot (m) redondant
Robot manipulator with more degrees of freedom than required for execution of the robot task.

rehabilitation robotics – Rehabilitationsrobotik (f) – robotique (f) de réhabilitation (f)
Robotic systems helping paralyzed persons or substituting lost motor function. Robotic systems can also execute training of paralyzed upper or lower extremities. Special mobile robots can guide blind people.

remote center compliance (RCC) device – nachgiebiges Werkzeug (n) – outil (m) compliant RCC
Passive device at the robot end-effector allowing small translational and rotational displacements which make part insertion operations easier.

repeatability – Wiederholgenauigkeit (f) – répétabilité (f)
Variance of robot end-point positions obtained during repeated movements performed under the same conditions.

resolver – Drehgeber (m) – résolveur (m)
Device converting rotational or translational velocities into analog electrical signals.

robot cell – Roboterzelle (f) – cellule (f) robotisée
Group of robots, workstations and transport systems in which a single family of parts is produced.

robotics – Robotik (f) – robotique (f)
Science of designing, building and applying robots.

robot learning – robotisches Lernen (n) – commande (f) de robot (m) par apprentissage (m)
Robot learning is performed either on-line by teach pendant or off-line through computer programming.

robot language – Programmiersprache (f), robotische – langage (m) de programmation (f) robotique
Computer programming language with commands enabling interaction between robot system and human operator. It is based either on robot movements or on robot tasks.

robot system – Robotersystem (m) – système (m) robotique
A robot system includes robot manipulator, power supply, control system, grippers and sensory systems required for the accomplishment of a robot task. A robot system comprises hardware and software.

roll, pitch, yaw – Rollwinkel (m), Nickwinkel (m), Gierwinkel (m) – roulis (m), tangage (m), lacet (m)
Three angles determining the orientation of an object in space.

rotation matrix – Rotationsmatrix (f) – matrice (f) de rotation (f)
3×3 matrix describes orientation of a coordinate frame with respect to the reference frame. It is also used to represent rotation.

rotational joint – Rotationsgelenk (n) – articulation (f) rotoïde
The rotational joint constrains the movement of two neighboring segments to rotation. The relative position of one segment with respect to the other is given by an angle of rotation around the joint axis.

S

SCARA robot – SCARA Roboter (m) – robot (m) SCARA
Selective compliant assembly robotic arm (SCARA) has two rotational and one translational joint. Its workspace is of cylindrical shape. SCARA robots are used predominantly in assembly processes.

sealing, robotic – Abdichtung (f), robotische – soudure (f) robotisée
Robot moves along the sealing path while applying a precise amount of sealing compound.

segment, robotic – Glied (n), robotisches – segment (m) de robot (m)
Robotic segment or link is a basic part of the robot mechanism connecting two neighboring joints.

sensor fusion – Sensorintegration (f) – fusion (f) de capteurs (m)
Integration of data from diverse sensors in the robot environment with the aim to produce reliable information required for operation of a robotic system.

service, robotic – Service (m), robotischer – robotique (f) de service (m)
Nonindustrial use of robots. Applications include health, safety, cleaning and maintenance, food delivery and entertainment.

shipbuilding, robotic – Schiffsbau (m), robotischer – construction (f) navale robotisée
Application of special robotic systems for welding and coating of large hull structures of ships.

simulation, robotic – Simulation (f), robotische – simulation (f) robotique
Robot simulation represents a useful computer tool in off-line robot programming and planning of robot cell actions in the virtual environment.

slip sensor – Schlupfsensor (m) – capteur (m) de glissement (m)
Sensor that measures distribution and amount of tangential component of the contact force in the robot gripper.

sorting, robotic – Sortieren (n), robotisches – tri (m) robotisé
Robotic and sensory system discriminating different types of items and classifying them into appropriate groups.

space robot – Weltraumroboter (m) – robot (m) spatial
Autonomous robot system performing geological or atmospheric investigations in space.

spherical robot – sphärischer Roboter (m) – robot (m) sphérique
Robot with two rotational and one translational degree of freedom resulting in a spherical workspace.

stiffness – Steifigkeit (f) – raideur (f)
The relation between the amount of contact force and displacement of compliant environment.

surgery, robotic – Chirurgie (f), robotische – robotique (f) chirurgicale
The application of robotic systems in planning and execution of endoscopic (inspection of the interior of the body) and minimally invasive surgical procedures. Surgical robotic systems make use of medical imaging and provide high accuracy and repeatability of operation.

T

teach pendant – Programmiergerät (n) – boîtier (m) de commande (f)
Portable hand-held device containing pushbuttons, switches and joy-sticks used for on-line programming and positioning of the robot end-effector.

telemanipulation – Telemanipulation (f) – télémanipulation (f)
Manipulation of objects by the help of teleoperation.

teleoperation – Teleoperation (f) – téléopération (f)
Remote control of robot manipulators in hazardous environments or in space.

tendon drive – Seilzug (m) – robot (m) à câbles (m)
Transmission system from motor to a remote mechanism via flexible cables and pulleys.

trajectory – Trajektorie (f) – trajectoire (f)
Set of points through which the robot passes during the task.

translational joint – Verschiebegelenk (n) – articulation (f) prismatique
The translational joint constrains the movement of two neighboring segments to movement along a line. The relative position of one segment with respect to the other is given by the distance along the joint axis.

U

ultrasonic sensor – Ultraschallsensor (m) – capteur (m) ultrasonique
Device measuring distance by emitting a narrow band pulse of sound and detecting the reflected sound.

unmanned air-vehicle, drone – Drohne (f) – drone (m)
Teleoperated flying mobile robots mostly in military applications.

V

vacuum gripper – Sauggreiffer (m) – pince (f) à aspiration (f)
Pneumatic device enabling attachment of objects by the use of vacuum pressure.

vision, computer – Computersehen (n) – vision (f) artificielle
Use of camera system and computer to assess, interpret and process visual information.

visual servoing – Sichtsteuerung (f) – asservissement (m) visuel
Use of computer vision to control the pose of the robot end-effector with respect to the environment.

W

welding, robotic – Schweißen (n), robotisches – soudage (m), robotique
Robot assisted spot, arc or laser welding is currently the largest application of industrial robots. Robots for spot or arc welding are capable of arbitrary positioning and orienting of welding gun in the dexterous robot workspace.

workspace, reachable, dexterous – Arbeitskreis (m), greifbar, gewandt – espace (m) de travail (m) accessible, dextre
Reachable workspace represents the set of points that can be reached by the robot end-point. Dexterous workspace is a part of the reachable workspace where each point can be reached with an arbitrary orientation of the end-effector.

wrist, robotic – Handgelenk (n), robotisches – poignet (m), robotique
Mechanical system between robot arm and gripper, usually with three rotational joints whose axes intersect at the same point.

Further reading

Craig JJ (2005) Introduction to Robotics – Mechanics and Control, Pearson Prentice Hall, Upper Saddle River
Hoshizaki J, Bopp E (1990) Robot Applications Design Manual, John Wiley & Sons, New York
McKerrow PJ (1991) Introduction to Robotics, Addison-Wesley Publishing Company, Sydney
Natale C (2003) Interaction Control of Robot Manipulators, Springer, Berlin
Nof SY (1999) Handbook of Industrial Robotics, John Wiley & Sons, New York
Sciavico L, Siciliano B (2002) Modeling and Control of Robot Manipulators, Springer, London
Spong MW, Hutchinson S, Vidyasagar M (2006) Robot Modeling and Control, John Wiley & Sons, New York
Tsai LW (1999) Robot Analysis: The Mechanics of Serial and Parallel Manipulators, John Wiley & Sons, New York
Xie M (2003), Fundamentals of Robotics – Linking Perception to Action, World Scientific, New Jersey

Index

CPSIA information can be obtained at www.ICGtesting.com
233786LV00011B/56/P

9 789048 137756